Steve Crum, President, ACE Communications

"From novice to pro, every scanner user will find this book an indispensable tool for the fullest understanding and enjoyment of this fascinating technology."

Radio Monitors of Maryland

"I'm amazed at the amount of information provided in just one book.... This fine publication is a must for the SERIOUS radio monitoring enthusiast."

Monitoring Times

"Eisenson conducts the proceedings with warmth and good humor, discussing the hobby as if he was sitting around with friends, swapping tales. Indeed, this is an enormous, if not eclectic, collection of information....plenty of great reading"

Paladin Press

"Radio scanner enthusiasts listen to cellular and cordless phones, police and military transmissions, industrial security lines, and other supposedly private communication just as casually as we overhear conversations on an elevator. This wry, cynical, and immensely entertaining book describes the scanner world."

American Survival Guide

"You can't miss....Comprehensively describes the hobby, the people in it, the equipment they use, and where they get the frequencies to which their scanners are tuned....includes the less public, in fact illegal, pastime of listening to cordless and cellular telephone conversations."

Popular Communications

"A well-rounded operating guide for scanner owners. Eisenson delves into all areas of the world of scanning....good material in this chunky book, all going together to comprise a manual that proves to be useful, knowledgeable, and readable."

The **FCC**
defines a scanner as a radio
receiver that can automatically
switch between four or more
channels, stopping when a
signal is detected.

SCANNERS
& Secret Frequencies

Henry L. Eisenson

PUBLISHING GROUP
San Diego, California

Ninth Printing: April 1998

Seventh Printing: October 1996

SCANNERS
&Secret Frequencies

Published by

INDEX PUBLISHING GROUP
3368 Governor Drive, Suite 273F
San Diego, California 92122

Copyright © 1993 by
INDEX PUBLISHING GROUP

ISBN 1-56866-038-3

Library of Congress Card Number 93-78786

Publisher's Cataloging in Publication
(Prepared by Quality Books Inc.)

Eisenson, Henry L.
  Scanners & secret frequencies / Henry L. Eisenson.
  p. cm. -- (The electronic underground series ; v. 3)
  Includes bibliographical references and index.
  Preassigned LCCN: 93-78786.
  ISBN 1-56866-038-3

  1. Radio--Monitoring receivers. 2. Eavesdropping. I. Title.
  II. Title: Scanners and secret frequencies. III. Series.
TK6564.M64E57 1993   621.384'19
                                              QBI93-935

Printed in the United States of America
9

Produced by SOS Printing, San Diego, CA

Acknowledgments...

This project was initially conceived as a brief utilitarian pamphlet, and with any publisher other than Index's Linton Vandiver, it would have reached that goal and been finished long, long ago – and 64 pages would have been far easier to produce than 320. Some persons don't know what a "producer" does in a movie project, and others are equally ignorant of a publisher's role. They're similar, and without the guidance and professionalism of that walking thesaurus and the benefit of his 30-year publishing experience, you would not be reading this. And in case you don't recognize the term, there is no "thesaurus" in Jurassic Park!

Sincere thanks go to a lot of people who helped with this book, and I will unfortunately offend most of them by omitting their names from this acknowledgment since I'm limited to one page, but there are a few who simply *must* be included. Steve Crum is the entrepreneur who founded Ace Communications, and he taught me a lot about the states of the arts. Joel Guskin explained scanner marketing by Tandy, and he and his colleagues were valuable critics. Norm Schrein ("Mr. Scanner" and President of the Bearcat Radio Club) helped also, as did Carol Ruth and John Clark of Radio Communications Monitoring Association (RCMA). And to the quite remarkable Bob Grove, of *Monitoring Times* and Grove Enterprises, and one of the hobby's best informed experts, I offer my thanks for helping me understand what drives this fascinating hobby – his insights and observations helped separate the wheat from the chaff.

Hats off to Derick W. Ovenall, N3EGR, and Mike Schulsinger, who did a fine job reporting errors in the early printings of this book. In fact, of the many thousands of readers, they were the only two who submitted errors for this printing (and their proposed corrections have been made). Can you find any of the errors that they missed?

Finally, I have joined the multitude that owes a debt to the scanner world's most knowledgeable and prolific technologist, Mr. Bill Cheek. His *World Scanner Report* is wonderful – may that fountain of inside information never stop flowing.

Dedicated to my most important assets –
my family and friends.

Henry Eisenson

"Yeah, Honey, I'm alone and no one can hear this. Now let me tell you where I'm hiding."

Jimmy Hoffa

THE EXPECTED DISCLAIMER

Why This Book?

This book was fun to write. Its purpose is to entertain and inform the reader, and not to prescribe illegal activities but to illuminate and then *pro*scribe them. Though it takes a few (well-deserved) shots at the laws of our land and at those who enact them, they are intended as critique, not libel. Underlying all of that is an honest attempt to educate, to dispel some myths, and to help the consumer make good decisions in a techno-complex and fast-changing market.

Above all, the book was intended for current and prospective scanner enthusiasts, to help them enjoy their hobby by providing a combination of technical insight, operational guidance, frequencies, information on the law, precautions, and various commentaries not assembled elsewhere.

Though this project is one book of a series, it may be unfair to give it any sort of "underground" label. Readers will find it to be mostly tutorial, but the remainder includes information that *someone* would rather was never published.

Errors

The author and publisher made every reasonable effort to validate the information presented in this book, but were scrupulous to avoid activities that violate the law. Therefore, there may be errors in the procedures by which a scanner is modified to receive cellular, but there was no *legal* way to check. Similarly, there was no test of the procedures for listening to cellular and cordless conversations.

And there are surely other errors of fact, for which both the author and publisher apologize. On the other hand, we all deserve a good gloat once in a while, and you'll love writing all those self-righteous letters!

Gentlemen do not read each other's mail...
Henry L. Stimson, 1929

Unless they can...
Henry L. Eisenson, 1994

CONTENTS

Robert Burns
(1786)

Epistle to a Young Friend:

**"Then gently scan your
brother man."**

FOREWORD

Bill Cheek
"Dr. Rigormortis"

A major revolution has been quietly mounting since the early 1980's, the roots of which are traceable to 1974 when the first programmable scanner, the MCP-1, was introduced by Tennelec. Clumsy as were the MCP-1 and others of its generation, major progress toward easily operated, *all-band* receivers had clearly commenced. Two distinct problems stifled the revolution in its infancy: VHF-UHF technology had not yet become your basic ho-hum garden-party topic, and *memory* was something largely reserved for the human mind and maybe some very expensive mainframe computers that required entire buildings to house and operate.

Scanning receivers, such as they were prior to the MCP-1, required relatively expensive quartz crystals, one for each desired frequency. Most scanners of the day had a capacity for between four and ten crystals (channels), and that was it. If you wanted more channels, you bought more crystals and eventually swapped them in and out by hand after the crystal bank was filled to capacity. Inasmuch as VHF-UHF technology wasn't cheap in those days, coupled with the additional investment in crystals, it is no wonder that scanning had long been relegated to the status of a cousin and distant sideline to Amateur Radio. Tennelec's MCP-1 was the harbinger of this revolution, but scanning experienced no real growth until the advent of low cost integrated circuit memory and mass produceable VHF-UHF designs as military innovations of the Vietnam era filtered down to consumer electronic levels in the early 1980's.

By 1987, scanning receivers had evolved to 400 channels of programmable memory, and VHF-UHF technology (25 MHz - 1300 MHz) had become old hat, rife with low cost surface mount technology and inexpensive ultra-modern, space age, design. Now, in 1993, there are no firm limits to the number of programmable memory channels in a scanner (1,000 factory-stock, 25,000 and up, easily retrofitted by casual hobbyists); nor to the limits of frequency coverage (100 kHz to 2 GHz). This revolution is enjoying a massive surge in momentum. Scanning has evolved into a major league leisure and business enterprise quite apart from any other radio service or interest. Conservatively estimated, more than a million scanning receivers are in use today, and the market appears to be growing at a rate of about 20% per year. Scanning is in the big time now, standing independently on its own, as both a big business and a big hobby!

Scanning has matured as a major market interest, not only in obvious areas of manufacturing and mass retail sales, but also in third party, aftermarket, and cottage industry products and services. Major hardware product areas include antennas, coaxial cables and fittings, wideband preamplifiers, frequency converters, computers and computer interfaces, electronic parts and materials, test equipment including voltmeters and frequency counters, retrofit circuits, and do-it-yourself kits for a wide variety of enhancements and upgrades. Major areas of software and services now available in this scanner aftermarket include books, periodicals, newsletters, frequency guides, and databases (*disk, CD-ROM, and printed*), computer programs for control and data acquisition as well as for decoding of RTTY, CW, FAX, and other digital data. There is even a large, active gray market in the scanner arena which seeks to develop and provide information, ways, and means to thwart encryption, digital encoding, circuit blockers, and other legal and hardware obstacles to the population's intense, growing desire to monitor the radio frequency spectrum from DC to Daylight! The scanner market and its aftermarkets in the United States alone could easily be turning $250 million a year now! Indications are that this phenomenon is not limited to this country. Europe is experiencing an astounding surge of interest in scanning!

While the hobby radio community may very well be the major consumer of scanner hardware and services, other significant consumers of scanners and scanner aftermarket products include business and governments, federal, state, and local. The news media are well known major users of scanners, but by no means are they exclusive. The Land Mobile and Cellular Radio Services are large consumers of scanner products. Sometimes, a $400 scanner adequately meets the same needs as does a $5,000 test instrument. Volunteer firemen, police, private investigators, paramedics, doctors, lawyers, engineers, field service, sales, and other people on the go have become deeply involved with scanning, but these professions are not the limit, they're just examples. The list goes on...

As economical and easy as it is to enter the *hobby* of scanning, it is a quite different and exciting challenge to become proficient at the *science* and *art* of scanning. Both a methodology and a modicum of education are necessary to become sufficiently skilled and adept to derive maximum benefit and reward for the effort. Amateurs and hobbyists traditionally have guided and educated each other for the past twenty years, but Henry Eisenson tosses his hat into the ring with *SCANNERS & SECRET FREQUENCIES* as the first comprehensive, no nonsense, *how-to* and *what-for* perspective to come from the space-age professional scene. Mr. Eisenson tells it like it is, with sound expertise and a light and humorous touch. You'll sense, without intimidation, an avant-garde technology as you refer to this important work time and time again. *SCANNERS & SECRET FREQUENCIES* will entertain, educate, and prepare you as The Adventure begins...

Bill Cheek

Author: *The Scanner Modification Handbooks*
The Ultimate Scanner
Editor/Publisher: *The World Scanner Report*
Columnist: Experimenter's Workshop, *Monitoring Times*
System Operator: The Hertzian Intercept BBS

Samuel Johnson
(1751)

Essay: *The Rambler*

"Curiosity is one of the permanent and certain characteristics of the vigorous mind."

INTRODUCTION

Human Nature

The major difference between "human" and "non-human" is intelligence, the most visible manifestation of which is communication. If we could not transfer information from one to another, we'd each have to learn every lesson from personal experience, and the maximum knowledge of any one person would be that accumulated in a single lifetime. Therefore, communication is more important than an opposable thumb, erect stance, or reduced body hair.

We acquire and process information constantly. It's human nature to learn, and the drive to do so (we call it "curiosity") is frustrated by impediments. The human race is unrelentingly curious – we instinctively collect information. Paradoxically, restrictions strengthen that innate drive and make the ultimate access of the data even more satisfying. In short, it is consistent with human nature to penetrate barriers, to listen.

To eavesdrop...

Was Francis Bacon really convinced that "man was designed by God as a persistently inquisitive animal?" When did Descartes remark that "curiosity is among mankind's most enduring virtues and deadliest vices?" And was it Winston Churchill who replied that "a man totally without virtue is indeed curious?" Such impeccable references make it easy to justify ears to the wall. So...

Keep this handy when you need to rationalize eavesdropping upon your neighbor's erotic phone calls with his girlfriend.

What This Book is About

This book is *not* about peeping toms peering over windowsills, or audio-toms with ears pressed to motel walls. It only briefly mentions President Truman's once supersecret National Security Agency, which is now the Supreme Scannist in the East, with ears pressed to *every* wall. There will eventually be many books about such techno-warriors who actually won WWIII, but this isn't one of them

Somewhere between the prurient-minded eavesdropper and the one whose mission it is to defend our country lies a recognized international hobby of scanning the radio spectrum for interesting communication. That's the world that this book describes, and the hobby itself is as diverse (and as expensive) as many others, and far more satisfying than most.

Once again, science and technology have multiplied human skill to new heights of achievement, because the high-tech scanner is replacing the low-tech glass-pressed-to-the-wall as the primary means of eavesdropping. Scanners are everywhere, but only the tip of that iceberg is visible to the casual observer, and then only when it makes the news by sinking some political or social Titanic. In England, a retired bank manager claimed to have used a scanner to record conversations between the Princess of Wales and a rather intense admirer. That made the news, and was responsible for a surge in scanner sales in the United Kingdom as old ladies turned from their twitching window curtains to a pair of earphones. Here in the United States, a political scandal was precipitated when a hobbyist recorded cellular telephone calls by a Virginia politician and gave them to his opponent. Scanners work...

Though the topic of "scanners" is only rarely newsworthy, technology is improving, the cost of a given set of functions is dropping steadily, the menu is huge, and sales are climbing. It's a growth industry and obviously a good investment, judging by manufacturers who pour money into new products, new software, new custom integrated circuits, and lots of advertising in both "insider" magazines and the popular press.

The scanner world is comprised of three key segments. The most visible is the hobbyist *him*self (because demographics suggest that more than 90% are male), and estimates range to more than 1,000,000 of them nationwide, with varying levels of commitment, understanding, and investment. The second segment is made up of scanner manufacturers and retailers, who advertise generally within the community so are not very visible to outsiders. The third group is truly invisible to the outside world – it is a gray market, wherein it is legal to make, sell, buy, and own the "products" of that group, but it is generally illegal or immoral to use them. That group makes accessories for scanners that permit reception of restricted frequencies, publishes lists of secret frequencies including those used by government agencies and law enforcement agencies, sells antennas tailored to the (illegal) interception of cellular telephone calls and baby monitors, and even offers computer software that facilitates anti-scanner privacy, including scramblers.

Of course, the availability of these resources has perhaps created a fourth element that is not properly part of the "hobby" but nevertheless buys and uses scanners: criminals, who may be very interested in nearby police activity, or who convert information illegally gained into further illegalities such as extortion or unlawful stock trading.

The scanner hobby is driven by a combination of economics and technology. There are manufacturers that exist solely to design and build scanners, and some retailers sell nothing but scanners and accessories.

A growing market justifies investments in technology to maintain or increase market share, in expanded advertising of such products, and in more shelf space for them. The scanner industry is economically complex as well as technologically challenging, and depends upon the most aggressive electronics in the industry.

The scanner hobbyist community is subdivided by commitment, investment, and understanding of the technology. The most casual member owns a $75 scanner that was built with the most-used frequencies of local emergency services already programmed. The most dedicated (and wealthy!) hobbyist owns a variety of scanners for pocket, table-top, and automotive use, plus an array of accessories, frequency lists, antennas, and reference books. He may attend meetings of similarly-inclined scannists, subscribe to periodicals that serve his hobby and the scanner industry, and spend many hours every day sifting the radio energy passing by. His home might look something like a porcupine, bristling with the latest in antenna technology. He may even design his vacations so as to travel to areas with interesting scanning opportunities.

There is specialization in scanning. Some hobbyists focus upon one area of interest such as law enforcement, railroad communication, military bases, or commercial aviation. Most hobbyists are more generalized, listening to anything that captures their interest at a given moment. As will be seen, private telephone calls are often *very* interesting.

Secret Frequencies

There are none.

If you own a scanner, it can find *anything* within its bandwidth and range if you're sufficiently patient. Or, you can purchase a frequency list, hire a search service, call the communications manager of the group you wish to monitor, buy aviation references, buy *Popular Communications, Monitoring Times, National Scanning Report,* or any of a host of other resources that make a business of listing frequencies that might interest you, and for guidance, depend upon the newsletters published by the various scanner clubs listed in Appendix 2.

In the spectrum, few secrets can be preserved once a properly equipped hobbyist determinedly sets out to uncover them.

On the other hand, some interesting frequencies cannot be *easily* discovered, and they're as secret as any. If every frequency in use were published (Freedom of Information Act, and so forth) then instead of "scanners" they'd be simple manually switched radios. After all, if we know where something is, why search?

The Law?

Is it legal to publish the frequencies used by law enforcement agencies? Of course it is, but a scanner-equipped burglar can learn that he's triggered a silent alarm as soon as a police car is dispatched to investigate. This book is not intended as a prescription for criminals, but the only difference between the hobbyist interested in police activity, and the burglar, is the use made of the information received.

Not even our Congress and state lawmakers have been coherent regarding the monitoring of private telephone calls, and today's confused legislation proves it. Some states make it illegal to monitor *cordless* telephone calls, and the federal government has made it illegal to monitor *cellular* calls, but neither provides mechanisms for enforcement, and there literally is no way to detect violation.

Until 1993, it was perfectly legal to make, sell, buy, and use scanners that could receive either cordless or cellular calls, or listen to baby monitors. The new legislation initially put limits on scanner design, and was expanded to ban the production, importation, or sale of downconverters and similar hardware, though there is currently no restriction upon home-made or kit-built units that would *allow* the scanner to receive trunked 800 MHz communications (and, of course, cellular). Of course, once that legal-to-own and legal-to-build downconverter is used to tune a cellular call, the law is broken.

A few minutes on the freeway will convince anyone that some laws are often abused or even ignored. Nothing is faster than a motorist's reflexes when he spots a highway patrol vehicle enter the freeway. We often break laws (have you ever jaywalked?), but somehow it's human nature to become angry when someone else does so. Many of us become furious to learn that [some] scannists listen to cellular phones, police, military, industrial security, cordless phones, baby monitors, and other supposedly private communication just as casually as they overhear conversations at a supermarket checkout.

There must be a *law!* There must be a way to control these incredible violations of privacy! We've got to *do something* about these lurking eavesdroppers! Will they be convicted? Can they be caught? Do the police know? Do they care? We'll see...

And, of course, if you can't lick 'em – *join 'em.*

MICROPHYSICS

Definitions

We have to make a stop at basic radio technology, because the reader can't move very far without a certain understanding of a few key concepts and terms. Sorry...

Frequency is a concept that is critical to any grasp of how scanners work and are exploited, therefore this book will provide definitions and examples of the term (and the knowledgeable reader is invited to skip each of them). On the other hand, "frequency" is often a difficult point to get across, and even skilled electronic engineers might benefit from noting the mistakes the author might make in presenting this term.

Here's the first try, and it's a classic from 9th grade physics textbooks: we'll drop yet *another* object into that pond... It will contribute some of its kinetic energy to the water, displacing it vertically, and the water will then oscillate as it seeks a new equilibrium. The ripples that emanate from the point of impact describe a series of oscillations from above the average surface level to below, as shown in *Figure 1*.

In that figure, **amplitude** relates to the strength of the signal, and to the amount of energy in it. The eye can more easily detect (see) ripples of greater wave height, or amplitude, just as the radio can more easily detect (tune) signals of greater amplitude. The more amplitude, the more deviation from zero and therefore the more energy. If the phenomenon is viewed instantaneously (frozen in time), when the amplitude *above* zero is the same as the amplitude *below* zero, the average doesn't change and there appears to be no net energy difference between "signal" and "no signal." The fact is that the phenomenon is *not* static. It is moving, and that movement creates a perceived and detectable signal as the waves pass a static point, because to a fixed detector (an eye, or an antenna) the level appears to regularly move up and then down, and it is that phenomenon that is detectable.

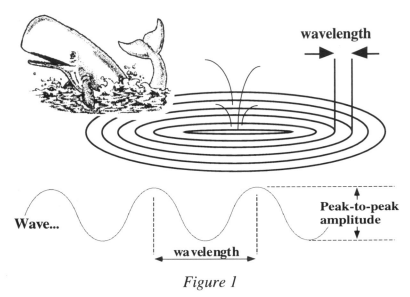

Figure 1

Visualizing the object dropped into the pond, it's easy to see that a certain number of those waves will pass a fixed point on the pond's surface in a given period of time, and the peaks of those waves will be some specific distance apart; that distance is called the **wavelength**.

The **frequency** of the phenomenon is the relationship between time and that wavelength. For instance, one might discover that five wave peaks pass a fixed reference point each second. When that is true, then the frequency of the surface perturbation is five cycles per second, or five Hertz (abbreviated "Hz"). If the peaks are one inch apart, then that's the **wavelength**. Don't forget that – there will be a test when you begin thinking of improving your scanner's reception.

Audio, Radio

There is, of course, a specific relationship between wavelength and frequency, and it's governed by the speed at which the energy travels through any given medium. If, at some ambient temperature, sound travels through air at a speed of 600 miles per hour, that's 880 feet per second. A tone at a frequency of 220 Hz (a mid-bass note within range of the male human voice) will therefore have a wavelength equal to 880' divided by 220, or four feet. A pipe organ designed to generate a 220 Hz fundamental tone will use a pipe that is one full wavelength (4'). Some organs can generate a true 32 Hz fundamental, which requires a pipe of nearly 28'. That math is easy because it involves sound, so we don't have to deal with rows of zeros, but radio travels at about 186,000 *miles per second*. If there were somehow a way to generate a radio signal at 32 Hz, it would require a "pipe" measuring over three million feet in length. Remember that when we begin considering antennas, because there's not a lot of difference between a pipe organ and an antenna, and they both use the same math.

Hum any note. Stop. You got those strange looks from the people around you when their eardrums were stimulated by alternating compression and rarefaction of air, caused when your vocal cords modulated the passage of air pushed past them by your diaphragm (a pump). Those alternations correspond perfectly to the ripples on the pond surface, and all sound can be visualized similarly. In fact, a loudspeaker generates sound because it is an electrically driven diaphragm that pushes and pulls the air mass of the environment, thus generating compressions and rarefactions. Those alternations of higher and then lower air pressure reach your ear, which acts like an antenna and focuses the acoustic energy onto the eardrum. That puts into motion the three smallest bones in your body, which then "tickle" nervous tissue that converts that movement into neural energy for your brain to perceive and interpret. In *Figure 2*, striking one drum causes the diaphragm of the other one to respond sympathetically even at a distance, showing that acoustic energy is indeed energy, and can do work. As you will see later, it does *more* when the second diaphragm is of such dimensions that it "resonates" with the first, like a tuning fork.

Regarding that note you generated, those bones are moving in perfect synchrony with your vocal cords (okay, you high-tech critics – we'll talk about *phase* some other time). You can safely say that they're both moving at the same frequency.

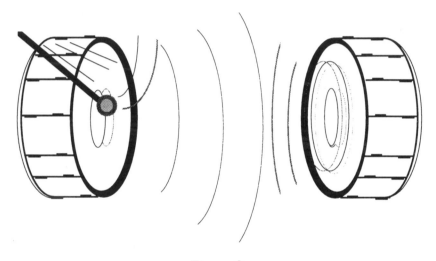

Figure 2

The odds are good that the tone you generated was at a frequency of about 220 Hz (90% of this book's readers are male, and that's a comfortable pitch for a man). That is, 220 sets of compressions and rarefactions passed a given point in one second. The lowest tone a human can hear is about 16 Hertz, and the highest is a bit over 20,000 Hertz (20 kHz), and none of us who have spent time in jet cockpits or discotheques can hear anything close to that.

As an aside, consider the relationship between frequency and perceived notes on the chromatic scale, a musical scale divided into half steps. An octave on that scale occurs with each doubling of frequency, so the range from 100 Hz to 200 Hz is an 8-note octave, as is the range from 1,000 Hz (1 kHz) to 2,000 (2 kHz), as is the range from 1 Hz to 2 Hz, etc. An octave is not a fixed number of Hertz, but begins with some reference point (the first number – a frequency in Hz) and ends at exactly twice that number. Now it's reasonable to ask whether anything can vibrate at a rate faster than the upper limit of our hearing. Things do – as any bat can prove.

So the range of possible audio frequencies increases, even to "ultrasound" where medical imaging instruments apply tones of millions of Hertz to the body, creating sound images of what lies within. There are audio frequencies in the megahertz range, and radio frequencies in the audio range, so it is not frequency that differentiates the two.

The difference between audio and radio is that audio signals mechanically move the medium in which they travel, and radio signals do something else. In a way, however, radio can start with vibration, and many of the rules learned for audio apply equally to radio energy. Radio energy cannot be demonstrated with two drums, but it's there nevertheless. Consider a microwave oven; radio energy, at some microwave frequency, moves the molecules of the material being cooked, adding energy to them (which becomes heat). But what makes radio energy operate at some specific frequency? In other words, how is it "tuned?"

When a quartz crystal is stimulated by electricity, it vibrates at some frequency determined mostly by its physical dimensions; the smaller it is the higher the frequency at which the vibration occurs. The vibration is extremely stable and pure, without distortion. A dime-sized quartz crystal might vibrate at a frequency of thousands of times per second. A tiny sliver (even smaller than those in quartz watches) might vibrate millions of times per second.

The upper vibration limit of quartz is generally less than 100 million Hertz, and only specialized crystals can reach that range. But radios operate at much higher frequencies, and benefit from the performance of a crystal. Fortunately for the author's ability to feed his children, there are technologies by which the stability of a crystal can be achieved at frequencies beyond the upper limit of quartz crystals.

The point is that the term "frequency" applies to phenomena that extend well beyond unaided human perception. It applies to bats, and to light.

It also applies to radio.

How Radio *Probably* Works

The computer engineer deals in information quanta that are either there or not, with nothing in between and no doubt. It's a square wave industry, with the signal level either at zero or at something, or above some point or below it. Digital design is a wonderfully deterministic profession that simply does not deal in shades of gray. The typical digital engineer tries hard to ignore the analog world, and suspects his sinewave counterparts of smoking cigarettes and eating red meat.

Analog engineering, on the other hand, is an art *and* a science. In analog design, not even the advanced-degreed engineer knows what's what until the last screw is tight and measurements are made. Twice. It's not black magic, but it's close. And the higher the frequency the likelier that mysteries will prevail throughout the design process. This is painfully true despite the existence of expensive computer simulation technology, which is digital and does a pretty good job of finding the area within the analog circuit where a problem might lie. Or might not. Radio-frequency (RF) engineering, or analog design, is not for the easily frustrated. With the advent of the wireless generation, the competent RF engineer will be valuable even if mysteries abound... in the land of the blind, even a one-eyed man can be king.

Radio is common, and is taken for granted worldwide, yet is a poorly understood physics phenomenon. The concept doesn't seem very complicated if you're willing to ignore a few minor mysteries, and accept the validity of a few mathematical expressions. A $10 radio kit can easily be built by a 15-year old, and sophisticated radios that listen to radio signals from the stars can cost millions of dollars and span hundreds of acres. There are literally hundreds of different types of radios, yet they all depend upon the same principles and mysteries.

Anyone can build a radio, but no one has a perfect understanding of how they work. Since Marconi first generated a few sparks that could be detected by a distant apparatus, radio waves have been deliberately produced and received by mankind. While there is some doubt as to what is really going on between the transmitter and the receiver, *something* is.

The communication link between the transmitter and the receiver works as well in a total vacuum as in air, can be established under water and through earth, and obeys laws well defined by mathematics. Yet no one understands what actually happens when a signal moves through the æther. That's a handy term that refers to whatever it is that the radio waves ride on or in, but we don't know what æther is, æther, and until we know what it is, we'll use the original spelling by which it was first hypothecated. Think of the æther as a fluid that can vibrate, or carry energy, but with zero viscosity – after all, planets don't give up energy to it or their orbits would decay and we'd fall back into the sun. We've been looking for the æther since 1880, and not even Michaelson (of interferometer fame) was successful in proving its existence. Obviously, he had no radio...

At any given instant on earth the æther is carrying millions of simultaneous signals from as many radio transmitters (plus incidental radiation from receiver circuitry), plus perhaps trillions more from extraterrestrial phenomena such as activity of stars, including that of our sun. The signals differ only in frequency and intensity (or loudness, which is the amplitude of the wave in the pond example, and the degree of pain caused by certain rock music).

The frequency range of ordinary AM broadcast radio is 550 kHz to 1650 kHz, and broadcast FM radio is from 88.0

million Hertz (megahertz, or MHz) to 108.0 MHz. Television, including cable TV, occupies several large segments of the frequency spectrum, and in fact the bulk of assigned frequencies below microwave are devoted to television. The highest assigned TV block is from 614 MHz to 806 MHz. Most satellite TV transmissions are in the range called "C-band" at about 5 billion (giga) Hertz, or 5 GHz, and others (in Ku-band) are at 12-14 GHz. Imagine! Something can actually vibrate more than ten billion times per second. Even Madonna... No. Tempting, but *no!*

Radar (which transmits radio energy into the distance and receives the bounced reflection from objects, etc.) can work at higher frequencies yet, and eventually one passes from the segment of the spectrum called "microwave" into "millimeter wave" and then into even less understood areas such as nuclear radiation, heat, light, and finally gravity (about which we understand almost nothing except that ignoring it often produces painful results).

The terms "frequency" and "wavelength" describe the same characteristic discussed in the example where we dropped an object (that object was a *radio*) into a still pond. A radio signal that operates at about 75 MHz oscillates seventy-five million times per second (75 million cycles) and travels at the speed of light, or roughly 186,000 miles per second.

Each "cycle" consists of a positive and a negative portion of the wave, and that complete cycle has a dimension, called wavelength, as shown in *Figure 3*. While frequency is described as cycles-per-second, or Hertz, wavelength is usually expressed in metric terms, hence "10-meter," "70-centimeter," and "2-cm" can all be translated to Hertz and back (since *cycles* per *second* has nothing metric about the units in the ratio). The constants are the speed of the moving phenomenon (radio and light move at ~186,000 miles per second), and then by the ratio of wavelength to frequency.

The concept of wavelength *(Figures 3 and 4)* must be accepted, if not entirely understood, since selection and operation of a radio antenna depends upon recognition of the relationship between frequency and wavelength (though many Pinto drivers have discovered that a wire coat hanger works just fine).

A receiver can tune any frequency within its range, and will detect any signal at that frequency. If only one signal is being transmitted at that frequency the communication will be clear, and if there are many the result will be muddied. At any given frequency there may well be dozens or even hundreds of transmissions around the earth, but a well-designed receiver will only detect the strongest of them. The receiver selects the desired signal using a filter of sorts, a selective circuit that is tuned to the frequency of the desired signal and rejects others.

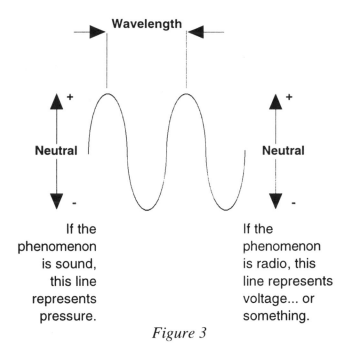

Figure 3

Radio spectrum is like shoreline. There's only so much of it, and it is therefore valuable because availability of spectrum

Radio energy moves at about 186,000 MPS (miles per second). Changes in voltage caused by a radio signal can be detected by a radio as the changes in energy level speed past some point – Beethoven's braille scanner, in this case. The rate at which those changes occur correlates with *frequency.*

Figure 4

defines our civilization's capacity to communicate, and to entertain its membership. Spectrum is a national resource, publicly owned, and controlled by government agencies. Modern transmitter designs conserve spectrum, partly because it's more efficient to broadcast precisely at the assigned frequency, without excess, and partly because the FCC forces such solutions because energy "spilled" into adjacent frequencies makes them unusable by others. Receivers, similarly, are designed to be receptive to only the tuned frequency and to reject others, simply because it's efficient to build them that way, though there are some designs that lock on to the strongest signal within its range, without tuning.

The spectrum of radio signals begins at audio frequencies (except that it's the æther that's carrying the signal rather than vibrating air) and goes on for many, many octaves. And remember, each octave ends at *twice* the starting frequency so "*doe*, a deer, a female deer," is at each end – an octave apart.

Let's start at 100 kHz, which is about the lower limit of the scanning hobby, and end at the beginning of microwave, which is about the upper limit of the best scanners commercially available. As can be seen in *Figure 5*, that's a span of about fifteen octaves – beyond the range of even a soprano ultima suprema... in pain. Each square represents one octave.

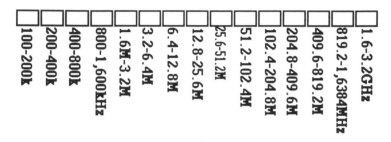

Figure 5

To understand some aspects of tuning a scanner, the concept of **harmonics** is useful. A harmonic is simply a tone, or frequency, at some multiple of the fundamental frequency (*n* octaves above), as shown in *Figure 6*. 2 MHz is the second harmonic of 1 MHz. 3 MHz is the 3rd harmonic, 7 MHz is the 7th, and so forth. That is important because the generation of a frequency used in tuning a radio usually generates one or more harmonic tones simultaneously.

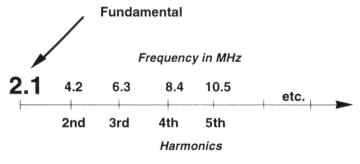

Figure 6

When a radio's tuning mechanism generates a harmonic, it can appear as a transmission by itself, or it can combine with other signals to generate still further signals, each of which the scanner erroneously recognizes as a transmission. Of course, designing tuning mechanisms that are relatively free from significant harmonics is an expensive process, but such tuners are less apt to lock onto non-existent transmissions.

Two of the most successful such designs are by Tandy: the PRO-43 and the PRO-2006, both of which use a "triple-conversion front end" that avoids the generation of undesired stray signals.

Tuning

How many of us remember thc days when a radio had to be "fine tuned" to produce good sound quality? Tuning was accomplished by rotating a variable capacitor wired in conjunction with a coil, or inductor. The combination "resonated" at some frequency, used by the rest of the circuit as a reference to select incoming signals.

That's the old way, and today there are few high quality radios (even in TV sets) that use the old analog tuning method shown in *Figure 7*.

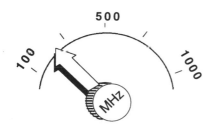

ANALOG tuning... The pointer can be rotated to *any* position, or value, between 100 MHz and 1000 MHz, and must be fine tuned to get best reception.

Figure 7

Crystal Tuning

Some radios are tuned with a crystal. One crystal is used for single-frequency operation, or several are combined with a switch to select among them (tune the radio). Obviously, such architectures limit the number of frequencies that can be tuned, and it's a lot of trouble to buy and insert new crystals when a channel must be changed. One of the finest FM tuners (even before the days of multiplex stereo) was by Karg; it used multiple crystals, and one ordered crystals for the local channels to be tuned. Early scanners also used just such a plan. It took a dedicated audiophile to own a Karg, and only a serious scanner hobbyist would take the trouble to manually replace crystals, yet such designs could often tune only ten channels or so.

There was an industry waiting to happen, and the missing ingredient was a way to retain the desirable performance characteristics of a crystal, yet permit convenient selection among multiple channels.

That "way" was found, and is the basis of modern radio tuning technology.

The Frequency Synthesizer

In those old receivers (up until the term "quartz lock" appeared on car radios) the input circuitry was effectively a filter that rejected undesired frequencies and passed the one to which the radio was "tuned." Today, most radios use frequency synthesis (*Figure 8*), and while there are two other kinds of synthesizer (direct-digital and direct-analog), the most common is "indirect," also referred to by terms such as "phase locked loop," "quartz lock," and "PLL." It is frequency synthesizer technology that allows the radio to tune precisely to any commanded frequency, with the accuracy, and lack of distortion, of a crystal.

A frequency synthesized radio can be identified by a digital display of the frequency to which the unit is tuned, while an "analog" radio might use a rotary or a "slide rule dial." The best user definition of a frequency synthesizer is that "when you set a frequency, it *snaps right in*." There's no ambiguity, no "nearly on the station but not quite" because the synthesizer is either *perfectly* on the desired frequency or *completely* off.

A key advantage of frequency synthesis is conservation of spectrum, since it allows tuning (of transmitters and of receivers) to be very precise. Precision means that little margin is required on either side of the desired frequency (ordinarily called "guard band") to prevent interference between channels, hence spectrum is conserved and more channels can be generated in a given segment of the spectrum, spaced more closely together.

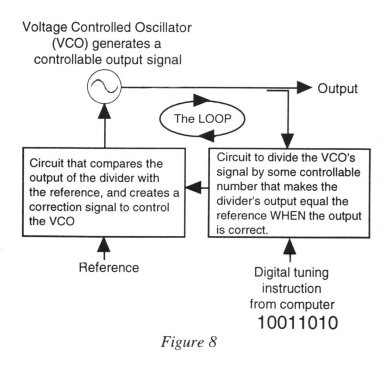

Figure 8

Accurate tuning across many octaves, whether by analog means or with a frequency synthesizer, is expensive and technically challenging. To make matters more difficult, the issue of "step size," or "frequency resolution" must be raised. A frequency step is something like a key of the piano; it's differentiated from the keys on either side of it by frequency, and there are no choices "between" the keys. A keyboard is digital. On the other hand, the piano tuner can slightly tighten or loosen a string to hit *any* note within its range. To continue the music analogy, think of a digital synthesizer as a musical instrument that cannot hit sharps or flats.

Wrist watches can be analog or digital. The hands of an old style mechanical "analog" watch move smoothly through 360°, so *any* time of day can be displayed by those hands because they pass absolutely every point in that circle. A digital watch, on the other hand, can display minutes, even seconds or hundredths of a second, depending upon how it's designed, but since it tells time in "steps," anything between those steps cannot be displayed.

Why is the difference between analog and digital important? Because it determines how much your scanner *cannot* tune. *Every* scanner uses digital tuning, and therefore has the potential of missing signals out there.

Let's say that some specialized tuner can reach a lower limit of ten million cycles per second, or ten megahertz, and an upper limit of 11 MHz. If it's an analog tuner, it can find literally any frequency between those limits, such as 10.0 MHz and 10.123456789 MHz, etc. A digital frequency synthesizer constructs its commanded output in steps of some dimension, so it cannot reach *any possible* frequency because some of them fall between those steps. In the example, the synthesizer might be able to cover the 1 MHz tuning range in 20 steps of 50,000 kHz each, or in one thousand steps of 1,000 Hertz each. That synthesizer could therefore tune to the frequency 10,111,000 Hz, but it would be unable to reach 10,111,500 Hz because it doesn't tune in sufficiently fine steps.

Another synthesizer design might tune in finer steps, covering that 1 MHz range in one million steps of 1 Hz each, but there would still be "spaces" between the synthesized frequencies that could not be tuned. Worse, the smaller the steps the more costly the design, and fine steps can bring other penalties as well. Step size is important, however. Examine *Figure 9*.

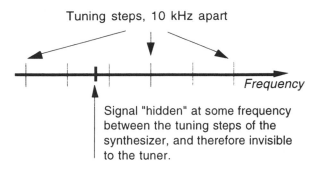

Tuning steps, 10 kHz apart

Frequency

Signal "hidden" at some frequency
between the tuning steps of the
synthesizer, and therefore invisible
to the tuner.

Figure 9

If your scanner can tune only to steps that are far apart (50
kHz, as an example), narrow-band signals can be missed and
the only hint that they are nearby *might* be a scratchy and
distorted noise. In *Figure 10*, if the unit is tuned to a
frequency as shown, with the indicated selectivity, it will also
receive signals at frequency "A."

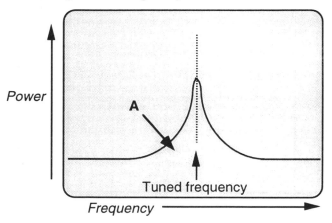

Power

A

Tuned frequency

Frequency

Analog tuning transmits power, or is receptive, at
frequencies other than the desired one. In this case,
receptivity includes not only the tuned frequency
but also the frequency at point A.

Figure 10

Compare *Figures 10* and *11*. In both curves, the total energy generated is about the same. Look at the highest peak, however, and it's clear that the synthesizer allows much more of the available transmitted energy (or receiver receptivity) to be focused exactly on the desired frequency while ignoring or rejecting other frequencies even close by, such as point "A".

Satellite television receive-only (TVRO) uses a small backyard dish antenna to receive signals from a satellite transmitter that is 23,000 miles above the earth. The satellite might use amplifiers with only seven watts or so, and the signal from them is distributed over most of the North American continent. How can a seven watt signal be received from 23,000 miles away?

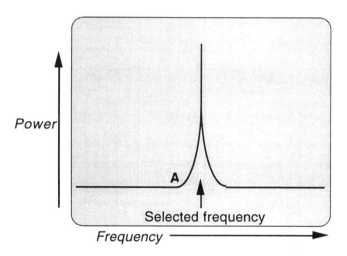

A *synthesizer* focuses power (in a transmitter) and receptivity (in a receiver) at the desired frequency, and wastes little on adjacent frequencies such as the one at point A. When less power is dissipated at nearby frequencies, more becomes available at the selected one.

Figure 11

Because the transmitter uses a very high gain directional antenna (which focuses available energy to the target, as you'll see later), and both it and the receiver use frequency synthesizers. This design creates a communication channel that acts almost like a tube connecting the two ends. As shown in *Figure 12*, nearly all the energy is expended exactly at the desired frequencies, and in the right direction.

One channel of the system might work without synthesizers if the transmitter were to broadcast enough power, and/or the receiver were wired to a dish the size of a city lot. TVRO has been possible for a very long time, but it's the advent of the modern frequency synthesizer that makes it – and high-tech radio – practical and economical.

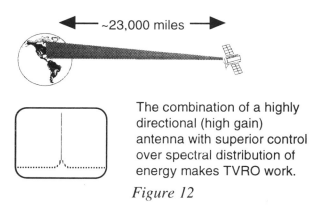

The combination of a highly directional (high gain) antenna with superior control over spectral distribution of energy makes TVRO work.

Figure 12

In a synthesizer, to achieve cost-effective performance we must accept compromises, and step size (frequency resolution) is usually one of the first factors to be sacrificed. The finer the steps, the more complex the synthesizer and therefore the more costly it becomes for a given level of performance. There are standards in the industry, however, which must be observed. Bands allocated to police usage, to broadcast entertainment, to cellular phones, and to military communications are divided into steps according to some nationwide or even international convention, otherwise a radio built for New York wouldn't work in Los Angeles.

For a scanner to find every signal, it must tune in steps no coarser than those used in any of the communication systems it is to scan. To be efficient, it should tune in the same steps. For instance, a radio that tunes in steps of 1 Hz can accurately find virtually any signal, but it would be very expensive and complex, and would take a long time to scan across several MHz. For scanning purposes, tuning steps of 1 kHz or even 5 kHz are usually sufficient, but for the hobbyist who wants to miss nothing, some very aggressive scanner designs support steps of only 50 Hz (AOR offers a good example).

Some inexpensive scanners have frequency resolution of 12.5 kHz and 25 kHz only, while better ones can tune in steps of 5, 10, 12.5, 25, and 30 kHz. Some scanners permit the user to select the step size, while others select it automatically based upon pre-programmed expectations in the frequency band of operation. For instance, when some scanners are tuned to cellular frequencies, the manufacturer's programming automatically selects 30 kHz steps, which is the channel spacing for cellular systems.

Later, we'll look at "channelization," which reflects the establishment of frequency steps for many portions of the spectrum. Some bands use steps 20 kHz apart and others, nearby, use steps twice as broad. These conventions and allocations were established long before the advent of frequency synthesis, hence they reflect the tuning errors or inaccuracies of archaic technologies.

Channel spacing is often wider (coarser) than technology requires. For instance, using advanced compression techniques, television channels could be placed closer together than they are, without hurting performance. Other portions of the spectrum are also used in inefficient manners.

As an example of allocation anachronisms, look at military aviation communication. Whether between aircraft or from air to ground, most such communication is between 225 MHz and 400 MHz, and channels are 25 kHz apart, as shown in *Figure 13*. This protocol was developed many years ago, before frequency synthesis evolved, but it perseveres today even though it is very inefficient. Speech on such channels occupies only about 3 kHz of bandwidth, and even allowing for drift of reference crystals, and "Doppler" (frequency changes as the transmitter and the receiver move toward or away from each other), 22 kHz is a lot of margin.

A properly synthesized 225-400 MHz voice system could probably operate with steps of 10 kHz, so 60% of the band could be returned to the FCC for re-allocation. But spectrum is precious, and the military is unlikely to give up one femtohertz of it (after milli, micro, nano, and pico, "femto" is a term reserved for the smallest things in the universe, such as the flight deck of an aircraft carrier at night).

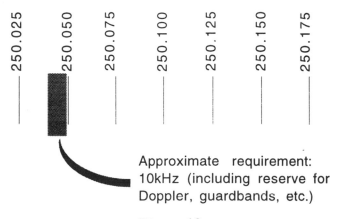

Allocation supports *N* bandwidth at 25 kHz steps, but at finer resolution only some fraction of *N* is required.

250.025 250.050 250.075 250.100 250.125 250.150 250.175

Approximate requirement: 10kHz (including reserve for Doppler, guardbands, etc.)

Figure 13

So to receive signals properly and efficiently, a receiver must employ a tuning apparatus that (1) reaches the desired operating range, (2) tunes in sufficiently fine steps so as to find the desired transmission, (3) has the ability to reject nearby signals while accepting the correct one, and (4) can accomplish these tasks without introducing distortion or false and distracting signals, or noise. A scanner is first a radio, and while it has a quite complex task list, it must do these fundamental "receiver" things before it can do anything else.

Switching Speed

Another parameter by which a synthesized radio is judged is switching speed, usually defined as the period beginning with the reception of an instruction to change, and ending when the unit reaches the new frequency. To understand this specification and its importance, take the matter to an extreme and consider how useless a scanner would be if it took ten seconds to tune each new frequency. Because virtually all scanners are tuned by a phase locked loop (PLL) synthesizer, switching speeds are usually in the range of a few thousandths of a second (millisecond).

Sensitivity and Selectivity

Sensitivity is an issue with any radio. In a real operational sense, the antenna is a part of the radio and must always be considered when discussing sensitivity of a system. On the other hand, the sensitivity of most radios is specified with a signal that is applied to the antenna input, and that information is used by the purchaser when comparing two models.

Today's radios are able to convert an antenna product of less than one half of one millionth of a volt (one-half microvolt) into a completely clear signal at the speaker. As a reference, consider that only 25 years ago a good quality stereo FM tuner required about ten times that level to produce a usable audio signal, and the FM radio operated over only a small fraction of the bandwidth covered by a scanner. Sensitivity is an important specification in scanners as in all radio receivers, and the development of advanced radio-frequency

(RF) amplifiers based on economical but ultra-low-noise transistors permits most scanners to show excellent performance in this category. Sensitivity alone isn't enough when there is more than one signal being transmitted. The ability of a radio to select between two signals that are close together is called *selectivity,* which primarily depends upon the *phase noise* of the synthesizer and associated circuitry.

In *Figure 14* the left example shows a low-amplitude signal that only a good receiver can differentiate from the noise. The right example shows that same signal but adjacent to a strong one. That poses a problem to the input circuitry, as the optimum circuit for tuning the low-level signal is *also* optimum for tuning the "edges" of the high-level signal. Creating two different and understandable signals under such circumstances is a technical challenge met only by well-designed (and usually expensive) circuitry.

Selectivity in Figure 14 is influenced by the amount of noise generated by the synthesizer at frequencies close to the commanded one. Technically, that corruption is called "phase noise" and is a critical factor in comparing one synthesizer with another, and one radio with another.

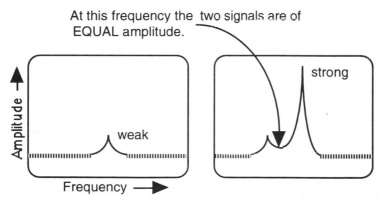

It's not difficult to develop a tuner that can detect weak signals. The difficulty is to make it selective, so a weak signal can be tuned even when it is close to a strong one.

Figure 14

As phase noise improves, more energy is focused exactly at the desired frequency, and less is wasted on nearby ones. Of course, in a receiver exactly the reverse is true, and the example applies to sensitivity. Visualize a finite amount of output power available and then consider what happens when all of it is concentrated at precisely one frequency and no other. Though it's not possible to achieve such performance, a "perfect" system would have infinite amplitude or sensitivity focused exclusively at the designated frequency, and no other, and therefore perfect selectivity as well. As the synthesizer approaches such perfection, so does performance of the overall radio.

Naturally, good performance in *any* of these parameters costs money, and in *all* of them costs a lot of money, and each factor seems to be inversely related to the others. As examples, fine steps are usually achieved at the expense of switching speed and signal purity, and broad bandwidth usually sacrifices sensitivity. Attaining excellence in all these parameters simultaneously is extremely difficult, requires complex circuitry, and is therefore expensive. *Very* expensive. A synthesizer that covers the range up to 1 GHz, with good performance, could cost more than $5,000, *plus* the reference which determines the stability/accuracy of the system.

Scanner Performance Factors

According to the FCC, a "scanner" is any radio that can automatically tune among four or more channels, stopping when a signal is detected. In any scanning radio, four performance factors are critical, and all are governed by the synthesizer. The first is *operating range*, or bandwidth, which determines that portion of the spectrum the system can reach and the highest and lowest frequencies that can be tuned. The second is *step size*, which defines how close together signals can be and still be tuned. The third is *switching speed*, which determines the rate at which the synthesizer can shift (scan) from one frequency to the next, and the next, etc. The fourth is *sensitivity*. There are other specifications, of course.

While phase noise of the synthesizer is important to the selectivity (and, often, sensitivity) of a scanner, spurious signals can degrade overall performance. A spurious signal, or "spur," is any energy in the output of the synthesizer that appears at one discrete frequency other than the one to which the unit was tuned. For instance, if the synthesizer is commanded to generate 100.000 MHz, and the output also includes a discrete signal at 200.000 MHz, that signal is spurious (uncommanded). In this case the spur happens to be the second harmonic, but spurs can be generated at almost any frequency.

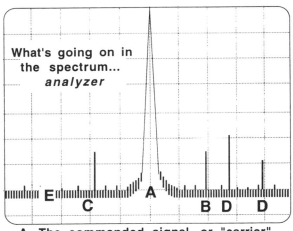

What's going on in the spectrum... *analyzer*

A   The commanded signal, or "carrier"
B   Second harmonic, twice the
      frequency of the carrier
C   A subharmonic
D   A spurious signal
E   The "noise"

Figure 15

Figure 15 shows a spectrum analyzer screen, which displays the distribution of energy across some span of frequencies.

That figure shows the commanded signal, harmonics, non-harmonic spurs, and "noise." For each commanded frequency output the synthesizer might generate a new pattern of spurious signals and harmonics, and the mechanisms that generate such unwanted discrete signals are sometimes so complex that the signals cannot be predicted. Sometimes, however, they can be either predicted or understood – and therefore ignored.

The problem with spurs is that each one shown in the display represents measurable energy at some undesired frequency. Spurs add and subtract, generating sum and difference signals at yet more new frequencies. And that energy is detectable.

A scanner tunes across the spectrum seeking faint radio signals from the antenna, but often spurious signals – and products of spurious signals as they combine in various ways in the radio circuitry – are sufficiently strong to convince the scanner that a valid signal has been detected.

The scanner's radio circuitry also generates unwanted signals, and though they're low in amplitude they're far, far closer to the antenna than, say, a taxicab downtown. The scanner doesn't know the difference between such a spur and a police transmitter, so after the scanner stops at the undesired frequency (called a "birdie" among hobbyists), the operator must re-activate the "scan" button, and/or lock out the undesired frequency. Some software supported accessories can sense and then reduce (or even eliminate) the birdie problem; the Commtronics HB-232 is a good example. Scanning is more productive when spurs are minimized, and good synthesizer design does that, but when other parameters are constant, the lower the spurs the higher the cost.

When the scanner's computer generates an instruction to "scan," the synthesizer control word defines the new frequency, the synthesizer shifts, any detectable signals are received, and the results are analyzed. If the computer decides that there is a signal of sufficient strength to remain at that frequency, squelch is defeated and the signal becomes audible. That process takes time. The less time taken by each step, the more channels (frequencies) the scanner can cover in a given period. Scanner speed is specified between 10 and 100 channels per second. At ten channels per second, 25 kHz apart, that's one-quarter MHz per second. Tuning across the military aviation band (225-400 MHz) looking for a signal would take a slow (10 ch/sec) scanner almost twelve minutes, while a fast one (100 ch/sec) would take only 1.2 minutes. Scanning speed *does* make a difference, though it's less important in a dense-traffic situation. On today's jammed freeways, the Corvette is no faster than a Pinto. Similarly, an ultra-fast scanner is no better than a slow one when every stop requires you to decide whether to push the "scan" button again.

A synthesizer that can cover many octaves of the spectrum in fine steps and with fast switching can be very expensive, especially when low phase noise and low spurs are important. The FS-5000 synthesizer, produced by Comstron (an Aeroflex company), is perhaps the world champ, and has become the heart and soul of many exotic radar, simulation, and electronic warfare systems. Depending on options, however, it can cost over $100,000, so it is *not* for your "quartz lock" car radio! It's easy to see why the scanner designer feels challenged – his goal is to achieve broad bandwidth, fine steps, and fast switching in a consumer product, and at consumer prices. It isn't easy.

Nonetheless, as you'll see, some modern scanner products are quite remarkable even by the standards of equipment in general use by military and government three-letter agencies only a few years ago.

Hardware Basics

A radio communication channel consists of three parts. A **transmitter** sends a signal, and a **receiver** detects it. Unfortunately, no one knows exactly what happens between these two, but it happens in the **æther** that carries the signal from one to the other. Don't move on too quickly... While it's true that there are mysteries about the æther, we have developed accurate mathematical models to define the effect of changing conditions (distance, antenna characteristics, power, etc.) and can use those models both to predict performance of a given system and to support the design of improvements. A serious scientist recently said that every time he turns on his radio, he is astonished that it works. It does, of course, but nevertheless there is little understanding of why.

While transmitters are necessary to the communication links that have attracted the hobbyist's interest, this book is about receivers, because that's what a scanner is. The rest of this book will usually ignore transmitters and instead will focus upon the scanner receiver. Some readers may think that this book has already covered more than enough physics, and they're beginning to flip pages looking for "secret frequencies." Be patient. The more you understand, the better you'll be at milking the most from your scanner. In fact we've still got some distance to travel down the path first found by Marconi.

When Marconi built one of the world's first "radios," the receiver was only able to determine whether or not a transmission was occurring, and no more, but that was sufficient. In that case, Marshall McLuhan was right. The radio energy (medium) *was* the message, and by turning it on and off in something of a code, information could be conveyed from transmitter to receiver. Today's receiver must do much more than simply turn on a light or make a click when it detects a transmission – in this era the light would never go out.

A modern radio transmitter accomplishes three tasks. First, it generates a carrier wave at the desired output frequency. Second, it adds the voice, music, or other information to the carrier wave ("modulates" the carrier). Third, it amplifies the combined signal and feeds it to the antenna, which then radiates the energy into the æther.

A modern radio *receiver* also has three jobs. First, it tunes to the desired point in the frequency spectrum where the transmission is expected, opening a "window" of sensitivity to a specific frequency (and the "size" of that "window" determines whether other frequencies slip through). Second, it amplifies the infinitesimal voltage generated when radio energy at that frequency passes the antenna, and amplifies again later in the chain to produce enough power to drive a speaker or headphones. Third, it detects and demodulates the communication that rode in on the transmitted signal. You see, a radio signal might "be there," yet *not* contain anything meaningful. While that's true for some talk shows, the statement is intended to mean that a radio can transmit a carrier wave (CW) only, without speech, data, or music added to it. In fact, CW radio is used to transmit dot-and-dash codes, including Morse, and under adverse conditions such primitive transmissions are often more penetrating than signals that include voice or other modulation.

To carry music or other meaningful material, the carrier wave is "modulated" by the transmitter. That is, the steady tone (CW) is changed in accordance with the material imposed on it. Visualize touching one key on an electric keyboard, thus creating a tone. That's like a carrier wave, and it's like what's being transmitted when you hear silence between records, or even between words, on a radio program. Now flip the switch that adds vibrato. The variations in the tone are minor fluctuations in frequency, like FM, which means frequency modulation. If you were to adjust volume rapidly, that would be like AM, or amplitude modulation.

Using FM or AM, programming is added to the carrier wave to produce a complex signal that is then detected and "demodulated" by the receiver. There are other ways to add intelligence/data to a carrier, but they don't apply to scanners so we'll pretend they're not there.

Receivers are specialized. Some can demodulate only FM, others only AM, but most scanner radios include both FM and AM receiver circuitry. In fact, a scanner must do many things well to compete on today's market because the signals to be detected are quite diverse.

A complete transmitter-receiver path might look like *Figures 16* and *17*. Note the process by which an audio signal is first amplified, then used to modulate a signal that is upconverted and tuned to the desired output frequency (by a synthesizer), and then the final signal is again amplified before it is passed to the antenna.

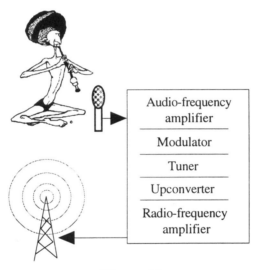

| Audio-frequency amplifier |
| Modulator |
| Tuner |
| Upconverter |
| Radio-frequency amplifier |

Figure 16

Figure 17 shows one possible receiver arrangement as the approximate reverse of the transmitter, but radio systems are usually not that simple, and scanners are among the most complex of radios. There's much more, of course, and each of the blocks in the diagram can be further expanded to show design detail down to the resistor level.

The typical hobbyist usually wants to know about controls, performance, and frequencies and is less interested in the specific design of, for instance, the downconverter block in a radio schematic. On the other hand, when problems such as undesired and useless "birdie" signals arise, a block diagram helps indicate the source and can help identify a solution.

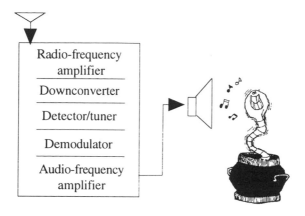

Figure 17

Despite complexity, difficult performance requirements, and serious economic and competitive pressures, today's scanners work well enough to be an effective bridge between the curious and the spectrum.

And sometimes, to a baby monitor in that bedroom across the street...

Simply Simplex

Some radio links are defined as "simplex," where everything happens at only one frequency. In a true single-frequency simplex system like that between an aircraft and an air controller, only one side can speak/transmit at a time. To use such systems, a transmission ends with the word "over," which tells the other party that it's time to switch between transmit and receive, and between listening and talking. Inconvenient, though it does slow down arguments. It's especially challenging to the scannist, who can usually hear only one side of the discussion and must infer the rest. Of course, if both the transmit frequencies are known, the scanner can be tuned to switch rapidly between them and nothing will be lost.

A more convenient arrangement is "two-frequency simplex," in which one frequency is used for talking/transmitting, and the other for receiving/listening, but they do not operate simultaneously. This is more efficient and supports a good argument, but the scanner can hear only one side of the fight unless both frequencies are programmed, as before.

A "full duplex" system is just like a telephone, and allows simultaneous screaming and name-calling at each end of the radio link. Again, two frequencies are used and the scanner must switch rapidly between them if the listener is to hear both sides of the contest. There are also repeater arrangements and radio frequency plans that allow scanners to hear both sides of a full duplex conversation on a single frequency.

Diverse Signals

Some interesting signals are barely detectable, and others appear at overwhelmingly high power levels, and they appear at any point from below 100 kHz to microwave. Some data and audio channels are less than 5 kHz apart and others are separated by 5 MHz or more. Some transmissions use amplitude modulation (AM), others use wideband frequency modulation (WFM) or narrowband FM (NFM), and many use single sideband (SSB).

To handle such a broad menu of diverse signals, a scanner must be a quite complex system. Successful designs are, and the complexity extends to very sophisticated custom integrated circuitry. The challenge met by a commercially viable product is to cover a useful and interesting portion of the spectrum, in steps fine enough to catch most of the communication out there, with such synthesizer performance that the signals are clean and differentiated one from another, and with such switching speed as to cover many steps in as little time as possible. No one of those standards is easy to achieve, and since they're usually negatively interrelated it's tough to attain a reasonable combination of them. It's even tougher to do it at a price consumers will pay, though that number is rising. For a tour through the frequencies, as you will see, consumers are happy to pay a surprising price in time and money – because the spectrum has become a fascinating playground. The entry fee is a radio...

But no radio can work until electrical energy appears at its input circuit. And that takes an antenna...

Charles Darwin
(1885)

**...damnable and
detestable curiosity..."**

THE ANTENNA 3

Another Mystery

The antenna is yet another mysterious element of a radio system, though science indeed has developed mathematical models that help us use antennas and match them to the task at hand. Even after 100 years of radio experience and evolution, there exists only a handful of true antenna experts, and each of them is so specialized that it would take all of them together to get one comprehensive authority on antenna design and performance. While this chapter surely contains errors of commission and omission, and probably violates some laws of physics, the experts who might best critique it are too busy theorizing to ever read this.

The objective of this text is to keep the scannist reading long enough to impart basic information on antenna performance and selection. So remember... this is *basic, practical* antenna information, only one step closer to physics than aluminum foil...

It's important to develop respect for the antenna since it can have such a profound influence upon the effectiveness of a scanner (or of any radio, for that matter), and antenna decisions must be made for the preferred frequency bands in a given geographic area. Fundamentally, all antennas perform the same task for the radio receiver by converting whatever is happening in the æther into an activity that can be amplified, tuned, and manipulated by electrical circuitry. To that extent, the antenna may be viewed as a device that takes energy from the aether and makes it "visible" to the radio. We may not understand it, but we certainly can use it. Therefore, we'll generally limit this discussion to the most lightweight mention of simple antennas, of which some are shown in *Figure 18*, and will spend only a bit more time on the antennas with which scanners are ordinarily sold. Because some enthusiasts wire their scanner to more complex antennas, they'll be mentioned and described, but not explained. We'll leave that to the very few competent antenna reference books in the Bibliography. Also, since the scanner is a receiver, the text omits those antenna factors critical to transmitter operation, such as power dissipation.

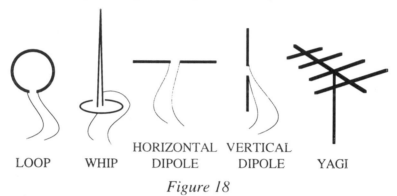

LOOP   WHIP   HORIZONTAL DIPOLE   VERTICAL DIPOLE   YAGI

Figure 18

Radio Signals

Radio signals don't always behave in intuitive ways, especially as the frequency goes either very high or very low. The lowest frequencies (below 500 kHz) penetrate the earth and large non-metallic structures, and are used for mapping caves, exploring pyramids, and for global communication with submarines. Some miles-long ultra-low frequency antennas are operated as low as 20 Hz (!) with properties as poorly understood as signals in the millimeter-wave bands up to 90 GHz.

Low frequencies below 30 MHz actually bend with the surface of the earth, following its contours and reaching the antenna of receiving radios even many hundreds or thousands of miles away. As the frequency goes up, the radio signal tends more and more to travel in a straight line, as shown in *Figure 19*. Because of that, higher frequency signals are referred to as "line of sight.". If a hill, a building, or our planet interferes, such radio links won't work.

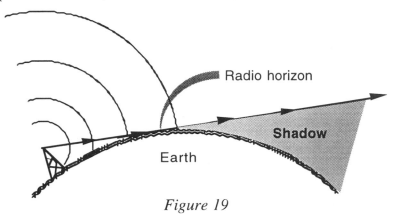

Figure 19

The range of a line-of-sight radio link depends upon the height of the transmitting antenna and the height of the receiving one. Obviously, if either is atop a tall building or a mountain, the range is extended. If the transmission is at a low enough frequency, or if there is a clear *line of sight* between the two antennas, radio communication is possible.

One may figure line of sight range by taking the square root of the antenna's height above the ground and multiplying it by 1.4 to find the distance in miles. For instance, if the bottom of an antenna is 9' above the ground, the square root is 3; multiply by 1.4 and the range in miles is 4.2 miles. If the "other" antenna is 25' above ground, it's line of sight range is 5' x 1.4, or 7 miles. These two stations can enjoy line of sight communication at ranges up to 4.2+7, or 11.1 miles. So, range is expressed as:

$$\text{Range in miles} = 1.4 \text{ times} \sqrt{\frac{\text{Antenna height}}{\text{in feet}}}$$

The nice thing about this simple mathematical relationship is that it applies to every antenna, all else being equal. But there's room in physics for considerable more complexity.

Basic Math *(Required reading for Scanners 101)*

Most antennas are some sort of resonant circuit. What is "resonance?" It's a natural periodicity, or recurring regular change, attributable to intrinsic mechanical or electrical characteristics (and that's not from a dictionary). A tuning fork of a certain dimension will resonate at one specific fundamental frequency, a pendulum prefers to swing at only one rate, a weight suspended by a spring will bounce at one rate, and most combinations of reactive devices (capacitors and inductors, which store and release electrical energy) resonate at some specific frequency also. For similar reasons, an antenna also resonates, and the quality of that resonance has a lot to do with the antenna's effectiveness.

If an electric signal can travel from one end of a wire (the simplest whip) to the other, and back, in the time period of one cycle of the RF frequency, as in *Figure 20*, resonance will occur. The electric signal moves at about the same speed as light, which is 983,573,087 feet per second, or approximately 186,000 miles per second. By convention, wavelength is expressed as λ, which – in miles – equals ~186,000 divided by the frequency in hertz.

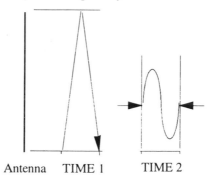

Antenna   TIME 1   TIME 2

To resonate, the half-wavelength whip antenna length must be such that radio energy can move from one end to the other, and back, in the time period of one RF cycle. TIME 1 must equal TIME 2.

Figure 20

Because the signal had to traverse the antenna twice, the length of wire necessary for the signal to travel a distance equal to λ in one cycle is λ/2, or one-half wavelength. Phew! That's what's meant by a half-wave antenna, and its success depends upon resonance similar to that which causes one drum diaphragm to *resonate* with air vibrations caused by the output of another. The same sort of calculation applies to quarter-wave antennas, 5/8 wave, and so forth, where only the maximum generated voltage varies with resonance, and therefore the length of the driven portion of the antenna.

Input Voltage

Resonance reduces the amount of work the radio energy must do to create a useful input voltage to the radio. At resonance, a clock's pendulum needs only a tiny push (energy) to keep it swinging. At resonance, the λ/2 antenna takes less from the passing ripples in the æther to drive the electrical input circuit of the radio, hence is more "sensitive." In *Figure 21*, the antenna produces a voltage with reference to the radio's ground, and that voltage is the input to the radio.

A typical scanner will operate well when the signal is at least a millionth of a volt (one microvolt, or 1.0 μV). That is, the antenna must convert the desired signal into about a microvolt. How much *power* is that? Things haven't changed since your last B+ in physics. It's not much.

Experts... forget SWR issues for a minute so the math can be kept simple. Calculate current first, using Ohm's Law (current equals voltage divided by resistance), so if the input amplifier of the radio presented a load to the antenna of, say, 100,000 ohms, then one volt across that resistance would produce a current flow of .00001 ampere, or 0.01 milliamp. Power is current times voltage, so in that case the power is 0.00001 times 1, or one hundredth of one thousandth of one watt, or 0.010 milliwatt. In fact, a good scanner produces reasonable audio performance with antenna output as low as a few millionths of a volt, which is still sufficient to drive the circuitry that follows. The antenna, therefore, collects the available radio energy (which isn't much) and points it in the right direction. Is the output of an antenna ever enough to cause a shock? In a thunderstorm, perhaps...

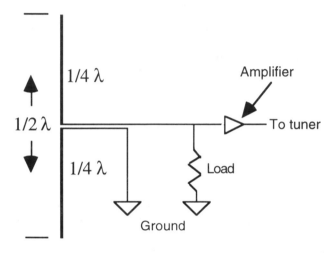

Half-wave resonance is best exploited by opening the circuit at the midpoint and then driving the two parts (in a transmitter) or feeding their output to a tuner.

Figure 21

The Antenna as a Filter

Consider the antenna as a resistor to some frequencies. Because its length determines the frequency to which it is most sensitive, it rejects (filters) other frequencies. *Figure 22* displays the performance of a simple "whip" antenna, equivalent to a straight wire. If such a simple antenna is subjected to a barrage of random radio transmissions, signals at some specific frequency (depending upon antenna length) will generate the highest input voltage to the tuner.

That means that there is one optimum length for each frequency, so at any other length the antenna will not be as efficient *at that frequency*, and at any other frequency the antenna will not be as efficient *at that length*. This is important information, and it also applies to all other types of radios. For instance, if you use a common adjustable dipole (rabbit ears) on your television, there is one length – and only one – that is best for each channel you tune. How many understand that to improve reception of a weak station one might have to *shorten* the antenna?

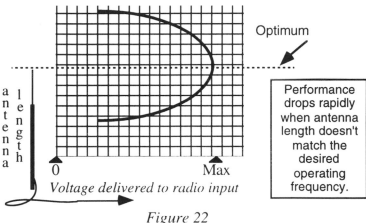

Figure 22

For scanners, which by definition operate over many frequencies, the best antenna is always a compromise. The "rubber ducky" antenna with which most handheld units are delivered is sometimes "loaded" by adding a coil to make it reasonably sensitive to (resonate at) a somewhat broader range of frequencies than a straight whip, but it won't be optimum to most of them. For utility and portability, a loaded antenna is best. For operation at a single frequency, or in a narrow frequency range, it is hard to do better than a piece of ordinary (stiff) wire of the correct length.

Perhaps the most useful and common antenna is the dipole, a vertical version of which was shown in *Figure 21*. It's like a television set's rabbit ears, but there's only one "ear" visible... In a handheld scanner, the length of the rubber ducky antenna is usually 1/4λ of the center frequency to be tuned, designed under the assumption that the user's hand and the body of the scanner will collectively serve the purpose of the other 1/4λ, effectively producing something like a half-wave antenna. But at what frequency? It's "half-wave" at one frequency, and something else at every other frequency, yet a scanner by definition tunes many frequencies. Also, one's hand might hold the scanner, or it might be resting against something else. All these variables have an effect that either adds or detracts from performance. Whenever the system deviates from an ideal dipole, it's efficiency that suffers.

And there's yet another factor that affects antenna operation. The *electrical* length of an antenna varies with its thickness. Viewed simply, the thicker the antenna the shorter it must be to present the correct electrical path to the signal. A relatively thick antenna will be shorter than the calculated length for a given frequency. This issue is critical only with exotic narrowband antenna systems, but mentioning it will save postage because all seven antenna gurus will surely send angrygrams if it's omitted.

A handheld scanner comes equipped with what can be viewed as a 1/4λ whip (the top half of the vertical dipole). A good length of such a whip (in feet) can be calculated by dividing 246 by the desired frequency of operation, in megahertz.

In a mobile or tabletop receiver, the antenna is usually a monopole, or whip. The vehicle or the tabletop radio becomes a "ground plane," which makes the antenna a whip. This throws the problem into a new category. Just like a dipole antenna, a whip is optimum at one frequency, and sensitivity will suffer at every other frequency.

Assuming a known frequency and an adjustable whip, one can calculate the length that will generate the highest input signal for the radio to further amplify and manipulate. That optimum length is one-half of the wavelength of the frequency tuned (which is why the reader was previously force-fed the relationship between wavelength and frequency). The correct length of a half wave whip (in *feet*) can be calculated by dividing 492 by the desired frequency of operation in MHz. You could have been taught that mathematical relationship long ago, but you'd always wonder why...

So, if you wish to determine how to adjust a whip antenna to receive military aircraft (225-400 MHz), you might set it for 300 MHz as follows: 492 ÷ 300 = 1.64 feet, and multiplying by 12 produces an answer of about 20 inches.

Suppose your objective is trunked communication at, say, 850 MHz. That's 492 ÷ 850 x 12 = 7 inches (or so). That specific dimension will work better than any other practical length. Probably...

Aren't antennas simple?

Signal Strength

In most areas, many signals of interest are repeated frequently, or transmitted from building tops and mountains at high power, so they are usually received at such a high signal strength that the antenna is not critical. Like a car radio in the downtown market, reception will be good whether the antenna is the right length or not.

When the scannist wants to listen to a relatively weak signal, or to differentiate between a weak one and a high power one, the antenna becomes important because it is the first filter in the circuit. The better it is at converting that signal's energy (and the energy of no other) to an amplifiable voltage, the better the performance of the system.

There are at least two ways to get more voltage out of an antenna. The first is to tune it for a specific frequency of interest by adjusting its length. Another way is to aim it like a gun. And when you point a gun at one target, it's *not* pointed at other ones. That's obvious with a Winchester, but with antennas the problem becomes a bit more subtle.

Omnidirectionality

Figure 23 looks straight down at a vertical dipole antenna (or any whip) mounted on a hand-held scanner, so the antenna is that dot upon the rectangle. The circle is a line that connects points of equal sensitivity. As can be seen, the antenna is equally sensitive to signals from every direction, and is therefore considered "omnidirectional." Assume a transmitter moved around the circle, through 360°.

An *omnidirectional* antenna will deliver the same voltage to the receiver regardless of the point on the circle from which the transmitter is operating.

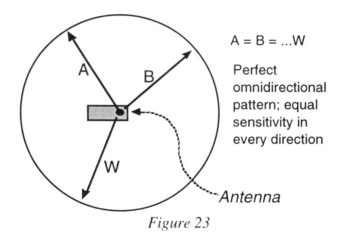

A = B = ...W

Perfect omnidirectional pattern; equal sensitivity in every direction

Antenna

Figure 23

Directional Antennas

A directional antenna sacrifices sensitivity in some directions in exchange for improved performance in other directions. *Figure 24* illustrates an antenna with such directionality, which can be described as "gain" just as an amplifier provides gain. Oriented as shown, the antenna has POSITIVE gain for (amplifies) signals emanating from points east and west, and NEGATIVE gain for (attenuates) signals from the north and south. Compare it to the omni antenna's sensitivity, and it is easy to see that to signals that appear from the front or behind, the directional antenna is far more sensitive than the omni design. In this analogy, the overall *area* within which a given transmitter will be received by the antennas is the same; that is, the *"total"* sensitivity of the two antennas is equal, but the second antenna will pick up signals from much farther away than the first, assuming only that the signal emanates from a point along the directional path of the antenna. That directionality makes the antenna an effective "amplifier" of those signals.

Without trying to cross the tutorial quicksand of antenna gain math, consider that *compared to an omnidirectional antenna,* a directional antenna has positive gain in some directions and negative gain (loss) in others, but there is a "sum" that doesn't change. Typical scanner omnidirectional antennas are so inefficient at some frequencies that they have negative gain. That's because the standard scanner antenna was designed to be reasonably sensitive across the frequency bands that might be scanned. If your scanner can work from 30 MHz to 1,000 MHz, it's easy to see that no single antenna can do a good job across the spectrum. In most cases, a simple piece of wire (of exactly the right length) will produce better performance at a desired operating frequency than the original design. To get sensitivity better than that provided by a simple whip or vertical dipole requires a sacrifice. A directional antenna *sacrifices* sensitivity in some directions to *gain* sensitivity to signals from others.

The simplest directional antenna is the horizontal dipole, as shown in *Figure 24*. It is somewhat directional, as the drawing indicates, is relatively insensitive to signals that appear from the sides, but is much more sensitive to front-rear signals than a whip or *vertical* dipole. The gain of a dipole is 2.2 dB (just a *little* math...), which doesn't sound like much until you recognize that the decibel (dB) reflects logarithmic relationships rather than linear ones, and a few dB can mean a lot. For instance, the difference between a 50W and a 100W light bulb is only 3 dB on the same scale...

The most common dipole is the TV antenna that brightens America's urban landscapes everywhere, though many are enhanced by passive elements to further improve directionality and therefore gain, and others use multiple driven elements (log-periodics – yet one more topic beyond the scope of this chapter). The traditional dipole "VHF" antenna is tuned to the broadcast TV bands in VHF, which coincidentally lie near amateur and industrial frequencies. The "UHF" antenna is sensitive to higher frequencies nearer trunking and cellular frequencies, and usually uses a separate (and different type) lead from the antenna into the home.

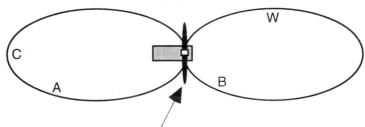

A directional antenna (dipole, in this case). The curves are defined by points of equal sensitivity. A transmitter with constant output power would produce the same input voltage to the radio from points A, C, B, W, etc.

Figure 24

The home TV antenna is quite directional, and produces better performance at many useful frequencies than any whip, especially if equipped with a rotator. A set of rabbit ears can also work well, even inside a house (assuming that the room does *not* have plaster walls on a wire lath). Matching the scanner to the typical twinlead and RG-59U cables that run from a TV antenna requires adapters available at any electronics parts store.

Since television signals are "horizontally polarized," and most signals to be scanned are "vertically polarized," the best performance with a television antenna will occur when the unit is rotated so the elements are up and down. If you see a rooftop television antenna rotated so the elements are 45° from the horizontal, it might be a loose mount, or a resident scannist using that antenna for both television and scanning.

There are also directional antennas that were developed specifically for scanner usage. Some focus on cordless phone frequencies, others on cellular, but the most useful are broadband and more generalized.

The Discone

Of the many different omnidirectional antennas for the scanner community, the best is perhaps the "discone," *(Figure 25)*, developed in 1945 as an inexpensive broadband antenna. The discone has zero gain (0 dbi), but that figure applies over a broad band of frequencies, which is its primary advantage because other antenna configurations lose energy rapidly as the operating frequency deviates from the optimum for that antenna (see Figure 22).

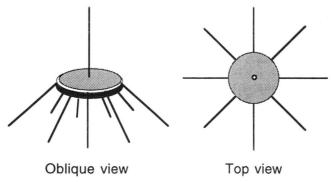

Oblique view   Top view
The DISCONE Antenna
Figure 25

The architecture includes a conductive disk (typically $1/4\lambda$ in diameter), plus angled radials extending as shown. There are many variations, of which the most common adds a whip that forms a separate 1/4 wave vertical antenna. Some units use four radials (spokes), others eight or more.

The discone antenna is equally sensitive over 360°, and has three other features useful to the scannist. First, performance versus frequency is fairly constant over (typically) three or more octaves in VHF and UHF, so it's a very effective broadband antenna that can receive almost anything. Second, it is far less expensive than other broadband arrays.

Third, it looks very complicated and impresses visitors.

For hand-held scanners, some companies make miniature discones, tuned to cellular/trunk frequencies. Ordinarily, the discone is much larger, and while it can be hung from the ceiling or put in the attic of a wood-frame or stucco home, it is best mounted outside, atop a mast. Like every book on radios, this one cautions the reader to be extremely careful when installing any antenna or mast around power lines, and to read and heed the instructions that come with the antenna and mast, particularly as they apply to lightning. There's no need to duplicate Ben Franklin's experiments.

Antenna Boosters, Pre-amplifiers

Many "boosters" are advertised, and their vendors claim substantial gain and improved sensitivity at prices ranging from $30 to more than $200. *All* are simple radio-frequency (RF) amplifiers – a function that already exists in the "front end" of a typical scanner. If a system only needed more amplification to achieve ultimate sensitivity, scanners would be built accordingly. But there's a limit to the benefit of amplification, particularly when price is a major issue.

Typical add-on devices amplify *everything* including noise, and for that reason are rarely useful. In such an approach, a booster is placed between the antenna and the scanner, as shown in *Figure 26*.

There *are* high-tech amplifying devices that "work," but you get what you pay for. There are a few buzzwords that might help avoid a bad purchase. A good RF amplifier will have a very low "noise figure," perhaps be built of a semiconductor material known as "gallium arsenide," use "FETs" or "Field Effect Transistors," and the circuit configuration might be defined as "parametric."

Consider a "preselector," a filter that blocks undesired frequencies so they won't be amplified and therefore interfere with the desired signal. Usually, an amplifier follows such a preselector circuit. A few manufacturers offer such combinations (Grove Enterprises is one), and they *do* work.

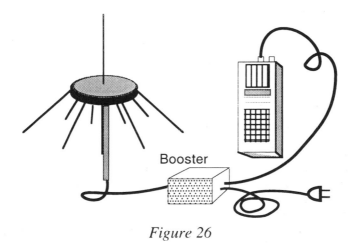

Booster

Figure 26

Since most third-party pre-amplifiers are sold by mail and therefore cannot be inspected or evaluated before the purchase, one of the best ways to avoid problems is to listen to the opinions of other hobbyists – yet another reason to belong to a club and/or read the periodicals, which print reviews of most available hardware.

One last point regarding amplifiers concerns spurious signals. No amplifier can increase power or voltage of a signal and produce an absolutely perfect image of the original waveform. Any deviation is called distortion, which consists of new frequencies in the output. These new signals are called "spurs" (uncommanded periodicities), and are generally harmonics, or multiples, of the original. They cause problems for the ensuing electronics because there is no electrical difference between a spur and a received signal, so "low spurs" and "low harmonics" are desirable traits of any amplifier in the RF chain.

An RF amplifier (particularly if it includes a preselector capability) that can truly improve the performance of a *good* scanner might cost more than some scanners. On the other hand, when you need it... *really* need it... there is nothing else that will do the job.

Reflection and Direction

Just when the young engineer begins thinking he knows something about antennas, up jump passive environmental influences.

Anything conductive can influence the propagation and reception of radio energy, in several ways. Though it looks much more complicated, a typical rooftop television antenna is basically a simple horizontal dipole wired to the television set. In addition to the mast and other supporting hardware, the many other metallic rods are there to "direct" or "reflect" radio energy to the dipole. They're not connected electrically, but they influence signal reception so much that adding those elements to the original dipole improves performance.

A scanner's antenna is sensitive to all metallic elements in the nearby environment, whether deliberately put there or not. Moving a scanner from place to place within an office building (metal studs, beams, desks, ceiling supports, other metal) will quickly demonstrate the effects of that sensitivity, particularly if the squelch is adjusted to keep reception barely "open." There will be points where reception is excellent, and others where the unit goes silent, and it's impossible to predict where reception will be good without trying it.

Even a lampstand within a few feet of a scanner might affect performance, and the effect can be either positive or negative. Add another conductor, such as the wiring within a wall a foot away, and performance can change once again. An antenna "system" really includes not only the obvious part, but everything nearby that can conduct, direct, or reflect electrical energy. It might include a human body, as well. We've all seen television reception change when we touch or approach a "rabbit-ear" antenna.

Just to make the system even more chaotic and even tougher to understand, physics adds something else to the equation – multipath. As the name implies, multipath occurs when the signal from the transmitter arrives at the antenna by more than one path. Typically, there's a single direct path, plus a major secondary one resulting from a reflection from a building, mountain, cloud, ionized layer in the atmosphere, etc. These signals arrive at the antenna at different points in phase.

What's phase? If two drums are beaten in exactly the same rhythm, at some locations the two sounds will reinforce, and at others they will interfere. That effect is due to phase, and it's equally applicable to radio as to audio. The result of multipath, then, is reduction of signal strength... or strengthening... or maybe nothing. *Good luck!*

It is literally impossible to calculate and then perfectly predict performance in an ordinary environment, because there are simply too many variables. One can operate a signal source and a scanner within an antenna range or Faraday room (spaces with all six sides made of grounded copper wire mesh, isolating the interior from the environment), and measure the results. That's how polar distribution of radiated energy is determined for different antenna configurations. But you want to use your scanner on the *other* side of the copper mesh, and once in the environment the system becomes a mystery even to the expert.

You can make a fetish about your antenna array, or stay practical and enjoy your hobby.

The Truth About Perfection

Be pragmatic. If a signal is "strong" at the reception site, it won't matter what antenna you use. Police communication, for instance, is so widely repeated (re-transmitted) in urban environments that an opened paper clip is usually sufficient. Anything more than a simple vertical dipole adds little to most casual listening situations, no matter what the advertising says. On the other hand, when you want to monitor the police in that town ten miles up the road, a modest directional might be just the ticket, but don't buy it till you've tried that old pair of rabbit ears from the garage or wired your rooftop antenna to the scanner.

In scanning, "perfection" is a state that occurs when you can clearly understand what's being said.

Compromises: Use What Works

Advertising works better than the products do, proved by the collections of antennas hobbyists have acquired over the years. Some make an improvement, but most do not. Collecting antennas is a boring hobby, so stick to the simple solutions.

The best broadband *portable* antenna (and generally the most convenient one) came with the unit – it's a vertical dipole or a whip. If you simply must replace it, the most sensitive practical portable antenna is a *collapsible* whip with some electrical mass below it (ground plane) with the length adjusted for a specific frequency of interest, though that requires a bit of math. For home or desktop installations, a ceiling-hung or mast-mounted discone is the best choice if the objective is to listen to signals from all directions, and a rooftop TV antenna if directionality and enhanced sensitivity are the goal. ARRL's book on antennas (see the Bibliography) is very useful, as it has a good blend of pragmatics and physics.

In meeting the antenna challenge, there are many paths into the realm of diminishing returns. It's possible to build a rotating mast with several band-specific antennas on it, to buy a broadband parametric gallium arsenide FET-based antenna booster with a preselector, or to pursue many other edge-of-the-art options. Like the audiophile who listens critically to the cable between the amplifier and the speakers rather than to the music, it's easy to waste money on antennas unless you are serious enough about scanning to have developed a system that makes small differences detectable.

That takes dedication, education, experience, and commitment.

But what's wrong with that?

George Bernard Shaw
(1933)

"An American has no sense of privacy. He does not know what it means. There is no such thing in that country."

SCANNERS

4

History

The scanner function originated when a sailor climbed the masthead to search the horizon for signal flags, but it's gotten much more complicated – and much more interesting – in the last century. It's also become less risky.

In both world wars, a scanner consisted of an array of complex equipment plus a well-trained human. Visualize a seaman aboard a destroyer eight decades ago. He's wearing earphones and slowly twisting a dial, tuning through the spectrum as he listens for a signal. When he hears something interesting, he rotates the antenna to determine the direction of the signal and to improve the sensitivity of his radio. If you had a glass or two of wine before beginning this chapter, perhaps you can imagine him reporting, "Captain, there's an SOS at three-one-zero degrees, very faint. I think might be the *Lusitania!*"

That sailor was turning knobs, but his hand plus the radio gear was the equivalent of an early scanner. The only difference between his gray-painted radio with a 6" dial and today's $200 hobbyist scanner is that the modern consumer product is wonderfully more automatic and effective. Add a directional antenna to a Radio Shack scanner and it can find transmitters that even the Gestapo would have missed.

Many years ago technologists responded to military requirements for automatic tuning and developed radios that swept across a designated portion of the spectrum, stopping when a signal was detected. It sounds simple enough, but in the days of vacuum tubes far more hardware was used to manage the system than actually received the signals.

Without doubt, it was government requirements (military plus three-letter agencies) that pressured the industry to develop fast-tuning, sensitive radios that could somehow detect, display, and eventually analyze all the radio energy picked up by the antenna.

In the last decade, scanning has moved firmly into the consumer market, and the result is a new and exciting industry. Though we cannot know much about classified government technologies, it is certain that today's hobbyist scanner is technically superior to scanners built for the National Security Agency not that many years ago.

Scanner Categories

Not that long ago, a doctor was a doctor and an engineer was an engineer, and that's all one needed to know. Today, thanks to specialization, one must ask a half dozen questions before you can determine just what kind of doctor or engineer is answering them. It's the same with scanners. The scanner specialty has subspecialized, there are many categories and variations on the market, and products have become divided by many physical, operational, performance, convenience, and cost factors.

Some scanners are sold with police, fire, and public safety frequencies permanently pre-programmed. The manufacturer uses an unchangeable memory that stores appropriate frequencies from around the country, and when the purchaser enters a code corresponding to his area, the scanner automatically shifts the frequencies for his area into its scanning instructions. Pre-programmed scanners might cost $75, or even less. They may be the simplest and least expensive way to enter the hobby, but interest fades quickly because the design restricts the scanner to those frequencies the manufacturer thinks may interest you, and there is no flexibility.

Though mail order houses sell that type of scanner in large volume, and certainly such designs are handy and a few specialized units are interesting, the remainder of this section will focus primarily upon *programmable* scanners. Such designs allow the user to program frequencies of interest into the memory of the scanner, which then sweeps through those frequencies one at a time. The memory might hold ten frequencies in elementary products, and 500 or more in advanced ones, and all may be programmed to match the interests of the user.

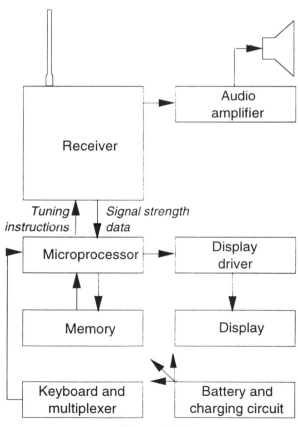

Figure 27

A block diagram of a typical programmable scanner looks like *Figure 27.* The logic (microprocessor/computer, memory, and perhaps the display driver and keypad multiplexer) are usually integrated into one or two chips, though collectively there may be tens of thousands of transistors within them.

Every scanner has a housing, keyboard or keypad, antenna connection, a power supply, memory, computer, audio amplifier, and speaker. Assuming the same computing power and quality of radio, the differences between various sorts of scanners are generally power, speaker size, and overall dimensions. The radio is, of course, the key function. *Figure 28* depicts a low-common-denominator "nucleus" radio receiver. Tuning and receiving functions are often embedded into one or two integrated circuits, with minimal supporting circuitry. The synthesizer is usually a separate chip.

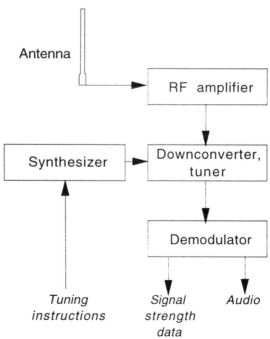

Figure 28

A single custom chip (called an Application-Specific Integrated Circuit, or ASIC) might do many simultaneous tasks, and might appear in many different products. In cases where a company makes several different products at several prices, special functions may be present in all models, but are only connected in the more expensive ones – the internal chip remains the same. It's analogous to early-generation cellular telephones, where all functions were present on all models of phones, but a glued-on cover prevented access to the extra keys on the lower-cost units.

Programmable scanners are, first of all, radios, with digital tuning that switches through frequencies as instructed by the computer. When a signal is detected, the radio tells the computer, which decides whether or not to stop scanning based upon programmed criteria.

General physical characteristics are determined by intended site of installation. There are pocket-size portables for use with internal batteries and a short rubber antenna, others for mobile (car, truck) use that take power from the vehicle and use an external antenna, and still larger units designed for table-top base operation, which take power from the wall socket and can be connected to an outdoor antenna. They all may be functionally similar, but they interface differently with the world. There are perhaps three categories of control features: basic, intermediate, and advanced. There are also a few exotic capabilities that appear on only a few designs.

As we discuss the current scanner industry, sodden with ultra high-tech ASICs and ultra low-noise GaAs FET amplification stages, never forget that perhaps the finest radio ever built (the Collins 390 series) used enough vacuum tubes to keep a room warm. The "latest" is not necessarily the "greatest." Scanners sold in 1992 were more capable than current units, as shown by used prices. Moreover, in the eyes of some experts, the finest scanner ever made was the Regency IIX-1000, but today this premium unit is generally found for sale only at swap meets and garage sales. Keep your eyes open.

Basic Scanners

The BC-70 variety has been a popular entry-level handheld unit for many years (Uniden/Bearcat).

Basic products include only mandatory control functions, and the display may show little more than frequency and/or channel number. Frequency coverage is spotty, often limited to narrow segments of the spectrum. The most commonly-encountered controls are shown in *Figure 29* (top) and *30* (front). Some basic units use an extendible rod antenna, but most permit detachable antennas via a connector known as "BNC," and have headphone jacks. This category of scanner may provide from 10 to 40 channels. Basic scanners have been fairly stable in price from 1989 through the present, and generally range from $80 to about $150 so cost little more than the preprogrammed variety.

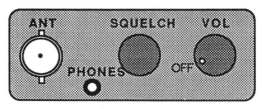

Figure 29

ON/OFF simply applies power to the unit. On many designs, that control is integrated with the VOLUME CONTROL.

The SQUELCH control is like an adjustable barrier to static. Adjusted too low and the radio sees static as a signal, so it stops scanning. Set too high, the squelch will block even desired local signals.

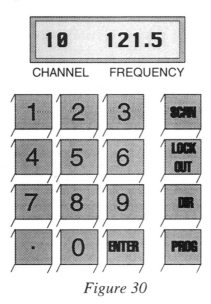

Figure 30

The SCAN button initiates scanning of selected channels, or of frequencies when in the limit scan mode. If the unit has locked on to an unwanted signal, pressing this button will resume scanning.

The DIR/DIRECT (sometimes MAN/MANUAL) button allows a frequency to be keyed into the unit if the button is preceded by depressing the program button. If the program button was *not* previously depressed, then this button will usually sequence through the programmed frequencies.

The PROGRAM button prepares the unit to load a new frequency into a memory channel.

The NUMBER buttons permit frequencies to be loaded into the unit. If the direct or manual button has been previously depressed, the unit tunes to the frequency entered. If the program button was depressed, the unit prepares to store the programmed frequency in memory.

A typical basic tabletop scanner, by Radio Shack.

The ENTER button tells the computer to react to the instruction just keyed in. If a direct frequency, the unit tunes to that point.

LOCKOUT tells the scanner *not* to stop at the tuned channel, though the frequency remains programmed and can be unlocked by pressing the same button again when that channel or frequency is tuned manually.

Many basic designs include a light, and a keylock to make the keyboard ineffective – thus safer to transport – by inserting either a mechanical or an electronic obstacle to keypad entries.

A few basic products include automatic Weather Alert detection, though for some marketing reasons this feature appears on only a few more expensive products. Broadcasts by the NOAA (National Oceanographic and Atmospheric Administration) include weather alerts initiated by a special signal to which such units are sensitive.

Intermediate Scanners

Scanners in the **intermediate** category include the preceding features (usually less NOAA alerts), though the MANUAL and DIRECT functions are separate. The "front end" will be more complex and may use higher quality amplification devices to improve sensitivity, and the frequency range will often extend through 800 MHz, though coverage still has gaps (usually including the cellular range) Such units allow the user to set a high and a low frequency limit, and will then scan through all frequencies between them. These scanners often include a "monitor" capability, which allows the user to store any frequencies detected while scanning between limits. There are other convenience and functionality differences that differentiate these units from the more basic ones, and intermediate units use a display that includes more information, as shown in *Figure 31*. Recent prices for scanners in this category range from $150 to about $250.

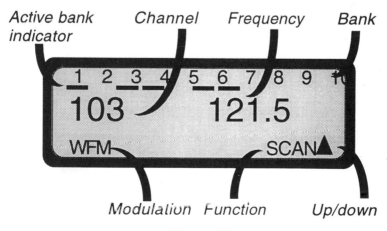

Figure 31

In addition to performance, additional functionality (buttons, or control functions), exists in the intermediate category.

MANUAL steps through the programmed channels.

DIRECT allows a frequency to be entered directly.

LIGHT illuminates the display, and sometimes the keys, usually for a pre-programmed period (a few seconds).

DELAY tells the unit to wait for a period (1-5 seconds) after a transmission ceases, with the expectation that the response will also be detected.

A typical intermediate scanner with a loyal following: the Uniden/Bearcat 200-series.

PRI, or PRIORITY, establishes one or more channel(s) as important. Scanners will scan, switch back to the PRI channel(s), scan again, return to check the PRI channel, and then scan, etc. Some designs allow a programmable period after which PRI channels will be checked.

CHANNEL BANKS are controlled by the numbered keys. In addition to entering frequencies, the number keys can select a bank of channels – ordinarily ten or twenty or even more per key. The "1" key might select channels 1-10, the "2" key channels 11-20, etc. Each of the channels is, of course, independently programmable. In most designs, the display includes a means of denoting which banks are "ON" (being scanned) and which are not.

MODULATION is controllable, usually by a button that toggles between AM and FM. Some intermediate units can demodulate both wideband FM (WFM) and narrowband FM (NFM) as well as AM.

The LIMIT button is used to set high and low limits for range scanning (scanning through all frequencies between two points, step size permitting). The user inputs the lowest frequency to be examined, and the highest, and then whenever the scan button is pushed the unit goes to the LIMIT SCAN mode and examines frequencies between the established limits.

The MONITOR button is used to store frequencies that the unit stops at when scanning between limits. It's like another memory bank, but instead of inputting the frequencies from the keys, they are input directly from the display driver.

Advanced Scanners

Advanced designs include features to enhance functionality and performance, and usually offer at least 200 channels in 20 banks or more. Very sophisticated and low noise RF front ends maximize sensitivity, and a higher performance frequency synthesizer further enhances performance.

Frequency coverage is usually continuous, and may range from 100 kHz to beyond 2 GHz. Switching speed (scanning rate) is very fast, since the logic (integrated circuitry) is selected for higher operating speed.

Advanced products cost from $250 to more than $1,300, and the best of them are functionally equivalent to those used by the government's three-letter agencies only a few years ago. An advanced scanner is also a near-equivalent to many serious shortwave receivers.

In most advanced scanners, STEP size can be user-defined. Pressing that button cycles through the options available, which are ordinarily 5, 10, 12.5, and 25 kHz. Most units that were designed to be easily modified to tune cellular frequencies will automatically shift to 30 kHz channel spacing when operating in the cellular band, but with the step size control that automatic feature can be overridden if desired. Some advanced designs support steps as fine as 50 Hz, which means that virtually no signal will be missed.

Selectable FILTERS are sometimes included, to permit noise attenuation and improved selectivity.

Multiple PRIORITY channels are common.

Advanced scanners can demodulate both NARROWBAND and WIDEBAND FM, as well as AM.

VOICE SQUELCH ("smart squelch") detects the difference between voice and noise, and resumes scanning if it determines that the signal being received is noise.

COMPUTER connections are included to permit spectrum display and expanded memory, plus software to support advanced functionality. In *Figure 32*, the computer might log "hits," provide spectrum displays, and more.

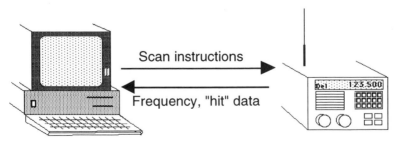

Figure 32

Some models offer EXPANDED MEMORY, with storage of more than 2000 channels, and as described in several publications (most notably in the *World Scanner Report*) some can be easily modified by adding memory to store 25,600 frequencies or even more.

Special VARIABLE OSCILLATORS (beat frequency oscillator, or BFO, and variable frequency oscillator, or VFO) permit special tuning techniques to separate signals that are close together.

The term "CTCSS" means Continuous Tone Coded Squelch System, and refers to a remote squelch control. That is, the scanner can be programmed so that a transmitted audio-frequency tone can instruct it to adjust squelch and "let in" a signal. Because such designs can be programmed to be sensitive only to a particular tone, it's almost like having a unique telephone number.

SINGLE SIDEBAND (SSB) reception is occasionally found in this scanner category; it allows detection of the most popular means of long-range communication. With high sensitivity, SSB capability, and a good antenna, the scanner becomes a serious short wave radio.

An *advanced* scanner: the PRO-2006 by Tandy. Though officially discontinued, very popular, and sold on the gray market at prices higher than retail of early 1993.

A higher-power AUDIO AMPLIFIER provides full-range compatibility with larger and more accurate speakers.

Some of the most aggressive advanced scanner designs include exotic features and functionality that few can exploit without formal technical training and expensive associated equipment.

An advanced tabletop scanner by AOR: the AR3000

Product Differentiation

The following table comprises a partial product listing derived from catalogs, advertising, observation in retail stores, and dialogs with hobbyists and dealers. It *includes* products that are customarily used as scanners, and *excludes* radios that are generally considered communications receivers even though many such products include a scan function. In many cases, the model suffix (i.e. "XLT") is ignored.

Prices are approximate and reflect typical discounts, not list or "suggested selling price." Uniden offers the Bearcat (BC) line, Radio Shack (Tandy) the PRO-XX/Realistic line, Cobra the SR series of scanning receivers, Ace Communications the AOR line, etc.. "Usage" refers to Vehicle, Hand, or Base station, though some units justify multiple notations. "Cat" refers to Basic, Intermediate, or Advanced. With two exceptions (the Cobra (uses frequency "UP/DOWN" keys) and the Trident, these units permit user programming through a keypad.

A number followed by "ch" indicates the number of programmable channels that can be stored in memory. The term "bands" applies to the number of separate programming bands into which frequencies can be programmed. The term "wx" indicates the ability to receive National Oceanographic and Atmospheric Administration (NOAA) weather broadcasts.

A number followed by "ch/sec" indicates scanning speed, in channels per second. In cases where a product is capable of more than one speed, the highest speed is shown.

A number followed by "μV" indicates input (antenna) voltage in microvolts required to receive a signal with reasonable quality. The lower the number (the more sensitive) the better, though the information comes from product advertising and is applicable to only one of the bands of which the unit is capable, and it could be worse in other bands. Note that sensitivity of the *receiver* has nothing to do with the antenna unless it is welded on and cannot be replaced by a better one.

Continuous coverage ("cont," no gaps over the scanning range) is noted, otherwise assume spot coverage of appropriate VHF and UHF segments. The highest tunable frequency, in megahertz, appears in parentheses.

This information was extracted from advertising, or was provided by manufacturers and distributors, and is subject to change without notice. Some discontinued units are listed because of popularity on the used and gray markets. Many products (particularly Uniden) are regularly and heavily discounted.

Model	Brand	Use	Cat	Comments	$$
AR900	AOR	H	I	100ch, 15ch/sec, .8μV, (950)	250
AR950		V/B	I	Like AR950, larger, (950)	290
AR1000		H	A	Broad bandwidth, cont, .35μV	450
AR1500		H	A+	500kHz-1.3GHz, SSB, .35μV, fast	500
AR2500		V/B	A+	2016ch, fast, cont, computer.35μV	500
AR2800		V/B	A+	1000ch, 500kHz-1.3GHz cont, BFO	450
AR3000		V/B	A+	100kHz-2GHz, 50Hz steps, .35μV	1100
BC60	Uniden	H	B	10ch, 10 bands, WX, (512)	180
BC80		H	I	50ch, 10 bands, WX, (956)	320
BC120		H	B	100ch, 11 bands, A/C, fast, (512)	260
BC144		B	B	16ch, 10 bands, .4μV, WX, (512)	160
BC147		B	B	16ch, 10 bands, .5μV, (512)	170
BC148		B	B	20ch, 15ch/sec, .4μV, WX, (512)	180
SC150		H	I	100ch, 12 bands, fast, (956)	350
BC178		B	B	100ch, 5 banks, fast, WX, (512)	240
BC200		H	I	200ch, 15ch/sec, (956)	240
BC220		H	I	200ch, 12 bands, WX, fast, (956)	440
BC350		I	B	50ch, 11 bands, fast, (512)	140
BC560		V/B	B	16ch, WX, (512)	170
BC700		V/B	I	50ch, fast, .7μV, WX, (956)	300
BC855		B	I	50ch, 5 or 15 ch/sec, .4μV, (956)	312
BC860		V/B	I	100ch, 15ch/sec, fast, WX, (956)	320
BC890		V/B	A	200ch, 100ch/sec, (956), VFO	440
BC2500		H	A	400ch, 1.3GHz, (1300), fast, VFO	420
BC8500		V	A	500ch, 25-1300, WX, VFO	660
BCT2		V	B	Hwy patr alarm + preprog scanner	270
BCT7		V	I	Hwy patr alarm, 100 ch, fast (956)	320

Model	Brand	Use	Cat	Comments	$$
DJ-series	Alinco	H	B	Basic scanner, A+ handheld	370[1]
MR8100	Uniden	V	I	100ch, fast, IBM interface, (956)	300
MVT7000	Yupiteru	H	A	200ch, fast, cont (1300)	400
MVT7100		H	A+	Cont, aggressive design	650[2]
MVT8000		V	A	200ch, fast, continuous (1300)	400
PRO23	Tandy/RS	H	B	50ch, (800), 25000 freqs	199
PRO24		H	B	16ch, WX, (956)	180
PRO25		H	I	100ch, (956)	250
PRO36		H	B	20ch, (512)	170
PRO39		H	A	200ch, 10 banks, fast, (960)	300
PRO41		H	B	10, (512)	120
PRO42		H	B	10ch, (512)	140
PRO43		H	A	200ch, 10 banks, fast, (999)	350
PRO44		H	B	50ch, slow, (512)	170
PRO46		H	I	100ch, 10 banks, (956)	230
PRO50		H	B	20ch, (512)	139
PRO51		H	I	200ch, (800)	300
PRO58		H	B	10ch, (512)	130
PRO59		H	B	8ch, slow, WX, (512)	100
PRO62		H	I	200ch, fast, (960)	300
PRO2006		V/B	A+	400ch, 26ch/sec, (1300) obsolete	400[3]
PRO2022		V	/B	A 200ch, fast, (960)	300
PRO2023		B	B	20ch, wx, (512)	160
PRO2024		B	I	60ch, (512)	180
PRO2025		V	B	16ch, (512)	130
PRO2026		V	I	100ch + preprogrammed, srch (956)	200
PRO2027		B	I	100ch, fast, search, (960)	230
PRO2028		B	B	50ch, WX, search, (512)	160
PRO2029		B	B	60ch, WX, search, (512)	180
PRO2030		B	B	80ch, WX, search, (956)	200
PRO2032		B	A	200ch, 10 banks, fast, search, (960)	300
PRO2035		B	A+	1000ch, 10 banks, fast, (1300)	450

Model	Brand	Use	Cat	Comments	$$
R1	Icom	H	I	100ch, cont, (1300), 2"x4"x1.4"	530
R7100		B	I	900ch, slow, continuous (1856)	1500
R10	Opto	H	B	FM: 30MHz-1GHz, specialized	360^4
SR001-B	Shinwa	M	A	IR remote, 200ch, 35ch/sec, cont	480
SR901	Cobra	B	B	10ch,.3μV, no keypad, (512)	75
TR-33WL	Trident	V	I	Hwy patr, radar, fast,.5μV, prepro	400^6

Notes:

1. The Alinco DJ series are handheld transceivers (HT), with a basic scanner function. They're considered good examples of that equipment category, occupied by quality transceivers from Kenwood, Yaesu, Icom, Standard, etc.

2. Yupiteru is an aggressively designed import with a short domestic history, but it is represented competently and therefore probably has a good future in this market when it can be redesigned to meet FCC standards. Until then, they cannot legally be sold in the United States. Some foreign retailers have advertised these products in the U.S. media, but reports have appeared that the product will be confiscated by customs and held in bond for the shipper to claim.

3. Several scanner retailers have bought the PRO43 and PRO2006 in bulk from independently-owned Radio Shack stores, and still resell them. Technically, these are both "obsolete" and are not usually retailed by Radio Shack. They are, however, competitive with any scanner product on the market today, and are frequently seen for sale on the "gray market." Some scannists have reported that a number of PRO43s were sold with nonmodifiable circuitry.

4. The Optoelectronics R10 is a broadband receiver with no frequency control. It locks onto *any* strong FM signal. Useful to, and perhaps originally intended for, the counter-intelligence market (which uses it to find "bugs"), it also permits hobbyist listening, but *always* to the strongest signal.

5. The BC2500 and 9500 are also triple conversion receivers, a tuning architecture that reduces or eliminates "birdies" and intermodulation products.

6. Several scanners carry the name "Trident," and Steve Crum is a co-inventor of the patented technique that loads pre-programmed law enforcement frequencies. Pre-programmed, yes, but one Trident model was included in our shopping list because it does so much else. It's a basic scanner (police, etc.), but this tiny package is also a radar detector, a laser detector, and, finally, it sounds an alert when it senses a radio repeater with which many highway patrol cars are equipped (1+ mile range).

The Crum patent? A state code and then a local code are entered, and buried in non-volatile memory are the frequencies for the entire country, from which the correct set is selected for the area of operation. It does not make coffee. According to Steve Crum, the combination of functions (scan/CB/radar detector/Highway Patrol detection), involves a design patented by Bobby Unser of racing fame. Trident products are manufactured by Ace Communications.

With his nephew AL, Bobby Unser holds a Trident scanner/radar detector/highway patrol detector that includes his patent.

That is a large menu, and it's not complete. How does the prospective scanner owner decide what to buy? Well, he can listen to the shopkeeper (who was instructed during that morning's sales meeting on what to "move"), read the advertising (written by unbiased expert consultants), or take time to seriously study the market. Few of us with the "wants" are willing to take time to do that, yet we all know what it's like to buy something and then a few days later wish the decision could be made again. Perhaps a guide to the *second* decision might make the first one a bit easier.

Hobbyists **replacing** *their first scanner* might select from the following:

AOR: Any

Bearcat: BC760, 2500, 8500

ICOM: IC-R1, 9000

Tandy: PRO-43 or 2006 (if one can be found)

Yupiteru: Any, and ditto

The Yupiteru MV-5000 handheld, a newly introduced advanced design.

Conversion Architectures

Some airlines prohibit the use of radios and other personal electronics because they radiate energy that can confuse the sensitive navigation and communication equipment on which the flight depends. Extraneous radiated energy is a problem to aircraft navigation equipment, and to other sensitive receivers such as scanners.

To convert a received frequency (as high as low microwave) to a frequency that can be easily processed and tuned, the received signal must be mixed with a tuning signal (from the synthesizer). The result is four different outputs: the original signal, the synthesizer output, the sum of the two and the difference between them, and only one of those (usually the sum) is desirable while the rest are not. A single conversion receiver produces many undesired signals ("birdies") that can deceive the logic (and annoy the user). A dual conversion receiver is more costly to design and produce, and is less likely to produce such spurious signals. A "triple conversion" receiver is the best choice of all, as a proper design can eliminate virtually all birdie problems.

A well designed triple conversion architecture can minimize such problems. Some triple conversion scanners arc:

AOR (all)

Bearcat 2500, 8500

Radio Shack PRO-43, PRO-2006 (good hunting!)

Yupiteru (all)

Tandy/Radio Shack

Tandy, the corporate operator of most Radio Shack stores, probably retails more scanners than any other firm in the world – all under the brand name "Realistic." Some "high end" products are manufactured by Imazeki, in Japan, and some "low end" units are privately labeled Uniden designs.

The 1994 Radio Shack catalog includes fourteen different scanners, and another dozen or so units hit the new catalog in September. The book also shows many accessories, antennas, frequency lists for police/ground action, an aeronautical frequency directory, and a marine frequency directory. Tandy sells its scanner products directly to the consumer, through more than 7,000 retail stores (that's the number in the catalog, but at least two sources claim that reality is ~9,000!). Some of those stores are owned by Tandy, and others are owned by private (franchise) businessmen who purchase goods from Tandy and benefit from corporate image and advertising, etc. That's an important distinction, as you will see.

The Radio Shack PRO-43: one of the best handhelds ever sold, and very compact. Newer units are similar in function, but not easily modified.

Tandy scanners range from entry level to near state-of-the-art, and are priced accordingly. At the high end, the Tandy PRO-2006 is the evolutionary result of many years of scanner experience and a major investment in marketing (determining what the customer wants). The 2004 and then the 2005 were for years the most expensive and feature-laden scanners on the market. During that period of dominance, Tandy learned a lot about what enthusiasts expect from "high end" scanners.

Tandy no longer sells (in the United States, at least) scanners that can receive cellular telephone calls, but some Tandy scanners will receive cordless phone conversations without modification. Historically, the better units (PRO-2004/56, PRO-43, etc.) have been easily modified to receive cellular frequencies. In fact, some believe that they were designed to be easily modified. One clue is that when the PRO-2006 was modified to unblock North American cellular frequencies, it automatically switches to 30 kHz channel spacing, the synthesizer step size for North American cellular. Of course, new legislation has changed the future of such products, and Tandy's plans for new ones.

And the Tandy flagship? According to the 1994 catalog the 2006 is still alive, though the PRO2035 boasts 1000 channels and has many 2006 features. The 2006 is fast, easily modified/improved, sensitive, has selectable step size and selectable modulation. While scanning between limits (a band defined by the high and a low frequency limits set by the listener) a monitor bank allows the unit to temporarily "remember" frequencies that seem interesting, and then scan through them or transfer them to permanent memory. It also has SOUND SQUELCH, which can differentiate between voice and noise. This design has more third party accessories and services available than any other scanner. It can be interfaced with a computer, remote controlled, and more.

Tandy/Radio Shack is a reliable retailer with excellent service and warranty, and that, combined with the superb technology that has gone into their scanners, makes it quite fair that they are near the top of the scanner pyramid, at least in volume. But that doesn't mean that Tandy remains technologically unchallenged.

ACE Communications

No one has pushed scanner technology as hard as Steve Crum. He was the VP/Marketing for Regency before it was absorbed by Uniden, and founded ACE in late 1986. That

Indiana company has grown to multi-million dollar proportions, and today ACE is an importer, a retailer, a wholesaler, a manufacturer, and a service facility. As an indicator of the technologies offered in ACE products, one of the firm's important customers is the National Security Agency, at Fort George G. Meade in Maryland, though the orders are generally placed by an innocuous group called the Maryland Procurement Agency. Nevertheless, it is certain that ACE products are used by three-letter agencies in the intelligence business, and it is easy to understand why.

ACE currently sells several scanner brands. From a performance viewpoint, the top of the list is probably the AOR product line, imported by ACE from Japan and sold both to consumers and to selected retailers. AOR produces several different models, ranging from high-intermediate to advanced+ in capability and functionality.

At $1300, the AR3000 is one of the most expensive hobbyist scanners on the market, but you get what you pay for. Though tiny, it is an extremely powerful receiver, and deserves the "best" rating given it by many serious users. It covers 100 kHz to 2036 MHz in selectable increments down to an incredible 50 Hz. That is, this design can tune to within 49 Hz of any frequency from below the broadcast band to low-microwave. It can demodulate AM, narrow- and wideband FM, and single sideband (SSB). The AR3000 is an exotic scanner, the advanced features of which will be useful only to experienced hobbyists. On the other hand, at the moment no one ever seeks to replace it with something "better." There isn't anything superior on the commercial scanner market, but there are a few ambitious competitors.

ACE identified another Japanese company, Yupiteru (anglicize this as "Jupiter") as moving up in technology, and acquired import rights to their products. Though the Yupiteru product line is currently limited to three models, more are coming, and ACE will exercise influence over their development. They are already very good, and evaluation of the new model 7100 handheld indicates that in the minds of some hobbyists it may challenge the most aggressive AOR products. Time and experience will tell.

ACE clearly intends to occupy the high end of the scanner business, and if there appears yet another pretender to the state of the art crown, the odds are good that it will be ACE that presents it to the buying public. In fact, the new Uniden offerings (8500 and 2500) are currently retailed by ACE.

Bearcat/Uniden

At one time most of the world's scanners were made under the "Bearcat" name by Electra Co., a division of Masco Corporation of Indiana, or by "Regency," whose name and technology were bought by Uniden (a Japanese company) and integrated under the Bearcat product line. The company once known as Regency is now said to operate as Relm Communications, in Florida, but they no longer manufacture scanners. Nearly all scanners are now made in the Orient.

Bearcat offers the broadest product line in the scanner industry, with designs ranging from primitive to advanced, though only recently this company announced products (the 8500 and 2500 XLT) for the most sophisticated users.

Uniden's new advanced entry – the BC8500

Bearcat scanners are sold by department stores, shop-by-mail catalogs, electronics superstores, television shopping channels, direct telemarketing, and other pipelines to the curious consumer. Uniden apparently builds and private-labels some of the Radio Shack scanners. Assuming that is so, and considering the remarkable distribution (variety of retail outlets) of those products, Uniden/Bearcat probably produces more scanners than anyone else.

While Ace plans to continue to dominate the high end of the scanner market, the new Bearcat 8500 and PRO2035 challenge the best that Tandy has to offer, and there may be further products introduced in the next few years since it is evident to Uniden that despite legislation, the scanner hobby is growing and consumers are demanding features and performance, and are willing to pay more for their perception of the best. And, there's always a market for the best of anything. Today, the "best" is made by someone else, but you can bet that the Uniden marketing people are working to define tomorrow's product, now.

And they have the resources to do it. Though specific figures are not publicly available, Uniden/Bearcat is the largest scanner manufacturer in the world, with products that – in one form or another – may have captured more than two-thirds of the total market.

Independent Specialists

The hobby magazines carry many of their ads. For years they've sold Radio Shack and Uniden scanners (usually high-end), often already modified. How can this happen?

Some privately owned Radio Shack stores sell scanners (and other products) to such third parties. Scanners are bought by the franchised store, resold in bulk (and at a discount) to the scanner specialists, who retail at a considerably lower price than Tandy charges over the counter. Third and fourth party access to Uniden/Bearcat products is less difficult, but the result is the same.

These specialists don't sell on street corners to passing yuppies in their Beemers, but some of them make it hard to tell the difference. They answer their phones with grunts and demand "Cashier's check, no credit cards, no personal checks, no CODs." And what about warranty? Radio Shack and Uniden both say that such modifications void the warranty, which makes service the responsibility of the specialist if you can find him when the unit breaks.

ICOM

The ICOM IC-R1 is an anomaly; it's both more and less than other scanners in its price range. ICOM is a Japanese company that makes very sophisticated equipment for the ham radio enthusiast, and the R-1 is in fact a complete *shortwave receiver* that easily fits in a shirt pocket. It is not as fast as other scanners and lacks some scanner features, but it offers several typically SWL features and is one of the very few with continuous coverage over such a broad frequency band.

Some hobbyists believe that the best scanners must be exotic ham communication receivers, though they are aimed primarily at the SWL (short-wave listener) market. The prices of these products might be measured in "Pintos," a trading unit equal to $500 if the car runs and the antenna is intact... A modest receiver might therefore be a two-Pinto investment.

Bob Grove, one of the industry's acknowledged experts, believes that the standard of comparison of cost-effective communication receivers is the ICOM R-7100, which can usually be bought for only three Pintos. It covers 25 MHz to 2 GHz, and while it lacks some scanner-specific features it is a fine radio.

The best communication receiver may be the ICOM 9000, which costs about ten Pintos, and with an optimum antenna array and appropriate accessories its price is almost one Pontiac. The 9000 has a front panel spectrum display that shows all signals in the selected segment of the spectrum, and it covers zero to 2 GHz. In terms of sensitivity, selectivity, and functionality, it is certainly one of the finest receivers ever built. On the other hand, it fits into your pocket only if you're a kangaroo. Nevertheless, it is an extraordinary product with fine performance and an outstanding human interface.

As a radio, or as a scanner, the ICOM 9000 – and a few others like it – are certainly near the best available to the consumer.

But what's the *absolute* best?...

Watkins-Johnson

As the manufacturer of compressive receivers (a highly sophisticated form of scanner) for the military and intelligence community, WJ has developed remarkable technologies that are only beginning to trickle down to consumer products. In terms of bandwidth, frequency steps, scanning speed, sensitivity, selectivity, and many other performance factors, the WJ receivers represent the most advanced technology possible today.

Few consumers would even consider such a radio, because a high-performance compressive receiver by WJ can be a twelve Pinto purchase. Add a first-class external synthesizer by Comstron and the price is measured in Cadillacs. But a WJ receiver *is* among the best-at-any-price, and some derivative of such equipment *can* be purchased by the consumer. So, surely, someone will. And considering the shrinking military budget, perhaps a reasonable place to start is an auction at a military logistics facility. You could get lucky.

On the other hand (and at the consumer level), WJ now offers communications receivers for the hobbyist (aimed at the Short Wave Listener, or SWL, market). They cost more than scanners, but their prices aren't measured in Houses...

Hardware Summary

There is always a market for the very best (and for the cheapest) of anything, and that's true in the scanner business.

If it were your task to market a primitive and inexpensive scanner, your promotion would focus upon what it *can* do rather than admit its deficiencies. The consumer must read between the lines and determine what each feature might mean in utility, and then what's important and what's not. Nevertheless, most first-time buyers of scanners replace their purchase soon after getting involved.

The entry-level consumer is therefore advised to buy intermediate or advanced products, to avoid preprogrammed scanners (except, perhaps, the special-purpose Trident), and to try hard to *use* a scanner for a while before *buying* one. As usual, you get what you pay for.

Scanning 101 – Getting Started

First, get your hands on a scanner, but keep your money in your pocket. Don't buy one unless you have no choice, because during your first few hours' experience you will probably define the features and functions you would like to have in the unit you purchase. If you buy one before using one, ensure that it can be returned or exchanged. The odds are against a hole-in-one on your first swing on a golf course.

Most of the control features on well-designed scanners are reasonably intuitive. That is, you can probably pick up such a unit and after an hour or so you'll have nearly everything working. None of the mistakes made during the learning process will produce smoke or damage the product, though a few units incorporate a control function that completely deletes all memory. A 400-channel scanner takes a long time to program, so when borrowing one with loaded memory, use caution. Though there are no written conventions in the scanner industry, most of the manufacturers seem to have converged upon similar control solutions. Here's a 2-minute step-by-step process to get noise from almost any scanner's speaker, though be warned that eventually you may have to (gasp!) read the instructions that came with it.

1. Using internal batteries or an external power cord, as required, turn the unit on. Ensure that the antenna is connected, and if there is a switch that selects between an external antenna and a built-on whip, switch it to the "internal," or "whip" position. Look for a button marked something like "keylock," and move it to the unlocked position. If there is a SPEAKER/PHONES switch, put it in the desired position.

2. Turn the SQUELCH fully counterclockwise. Adjust the VOLUME until the speaker sounds like a conch shell held to the ear on a windy day. Turn the squelch clockwise, and the moment the surf dies away, stop turning.

3. Press the SCAN button and observe the display, which should be scanning through whatever channels are in memory. Even units with selectable banks of frequencies require that at least one bank be active at any time. When a signal is detected and the unit stops scanning, adjust the volume. If no signal is detected but you think there should be one, turn the squelch counterclockwise and set it to the point where the static is barely cut off.

4. To change the status of any channel bank (from ON to OFF, for instance) press the key corresponding to that bank. The display will show a bar (see *Figure 31*) for each of the frequency banks that is turned ON.

5. To shift to a specific channel, press MANUAL, the channel number, and MANUAL again.

6. To tune a specific frequency, press DIRECT and then key in the frequency. With some units, you must press ENTER. A few hours' experience with a reasonably competent scanner can be a revelation. It can also be quite frustrating, so when all else fails... read the instructions! Even if none came with the (possibly borrowed) scanner, a phone call to a cooperative major dealer might produce instructions on a handy fax machine.

Let's assume that you enjoy your "trial," and eventually you actually pay money to own one of these units. The next step is to become dissatisfied, because after a few days or weeks, not even murder, riots, high-speed chases, and searches for the Amazonian Fruit Fly are enough. You want more. Much more. You want it *all!* Those "gaps" in frequency coverage must *go!*

You want faster scanning, more frequencies, more features, more memory, and a longer antenna.

Every man wants a longer antenna.

Franz Kafka
(1926)

"You do not need to leave your room. Remain sitting at your table and listen... The world will freely offer itself to you to be unmasked, it has no choice, it will roll in ecstasy at your feet."

MODIFYING SCANNERS

5

How to Fail

Scanners are housed in durable plastic and metal cases that deceive some into believing it might be difficult to damage the "works." It's not tough if you are persistent and have carefully studied the unit's vulnerabilities, so this chapter provides a few hints on how you, too, can use ordinary tools and Heathkit skills to utterly destroy a scanner... With a little time and a big enough soldering iron, almost anyone can disable one or more functions, or even demolish a scanner completely. Here's how:

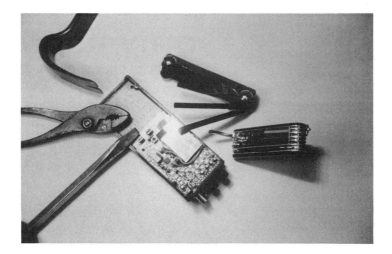

Basic weaponry for the scanner enthusiast. Only the gasoline blowtorch is missing.

1. Overheat a component. Every part within the scanner was selected by the manufacturer based upon its performance specifications and its tolerance of temperature. Plastic can be melted or a trace (the copper foil "printed" on a printed circuit board) de-bonded from the board. It's less obvious that an integrated circuit can easily be damaged by excess heat, and the problem is invisible. Many hobbyists use solder guns, but most such weapon can deliver enough energy to an integrated circuit to damage internal transistor junctions or bonds from the transistors to the outside case. To a low-mass item such as an IC, there's no difference between a solder gun and a propane torch, so all soldering should be done with a low-power, fine-tipped soldering iron. Another easy way to overheat a component is to carelessly use braided solder-removal tape. First heat the tape and *then* apply tape and heat to the solder to be removed. When working on *anything* electronic – remember: heat is the enemy of reliability.

2. Physically stress a component. The value of a resistor or capacitor can change if it is physically distorted/bent/twisted, and the advent of surface-mounted parts makes it easy to do. If the circuit board is even very slightly bent (in either direction) when the solder holding a part is melted, when the solder hardens and the board is relaxed, the part can be stressed and either fail or change value. Another way to physically damage a part is to melt the solder at one end and pry it from the board, bending the other end. If that's done just right, one can even lift the printed circuit trace from the underlying circuit board, ruining not only the component but the board itself.

3. Apply a static discharge. Yes, it's exactly the same sort of event that you experience when you shuffle your feet on the carpet and then touch a doorknob, but to most electronic devices that spark can be quite deadly. Electrostatic shock damage is the most insidious mode of failure in electronics today, and applied to the right spot, one tiny spark can destroy integrated circuits throughout a scanner. Most scanners use CMOS (complementary metal oxide

semiconductor) parts because they dissipate little power and
let batteries last longer, and since it doesn't make sense to
build one scanner "engine" for handheld units and another for
use on the tabletop, CMOS is everywhere, but it is
particularly vulnerable to static damage. Photomicrographs
have demonstrated that a static discharge can "pit," or
physically damage, the incredibly fine circuitry inside the
chips, but that damage isn't evident for months, or even years,
after the static discharge, when the unit fails for no other
apparent reason. If you're not into the sport of killing CMOS,
there are precautions you can take. High humidity lets the
static energy dissipate harmlessly, but you needn't wait for a
foggy day to do the job. Just be careful.

In general, one should NEVER open a scanner unless
precautions are taken. The easiest and simplest is to attach
(perhaps wrap?) a copper wire at the base of the antenna
connector – which is a good ground – and then wrap the other
end of that same wire around a metal wrist watch band or a
ring. That reduces the likelihood of an electrical voltage
differential between your probing fingers and the electrical
circuitry. Along that line, it's probably a bad idea to clean the
scanner case by rubbing it on rabbit fur.

4. Let a few strands of wire fall into the "works." It's like
Russian roulette, but less exciting unless you're an integrated
circuit. When you snip the end of a piece of stranded wire,
hold it directly over the circuit board and let the pieces fall,
hoping they'll find their way into invisible spots. That way
you can be pretty certain that *something* interesting will
happen, but you won't know exactly *what* until you turn on
the unit and hear either ET phoning home – or nothing.

*5. Leave the batteries in, or the power on, while working on
the unit.* This can get downright exciting, and can take out
not only the scanner but the hobbyist as well. A scanner
battery pack many not generate a dangerous voltage, but
there is probably enough energy in it to weld a screwdriver to
sheet metal, or to produce a serious burn. A tabletop unit that

plugs directly into the wall has deadly voltages present. If you decide to work with your unit "live," have someone with a video camera standing by.

6. *Apply the power backwards.* That's a frequent thrill in this "batteries not included" era, but most well-designed scanners use a "diode" (a device that lets current flow in only one direction) in series with the battery circuit to keep you from destroying polarity-sensitive components. But suppose the modification process requires a bit of electrical disassembly? Some connectors are "unpolarized" (they can be connected in more than one way), and if you don't reconnect them properly you can bypass the protection of that diode. The trick here is to avoid making sketches of the assembly or marking connectors, so that when you encounter ambiguity regarding a connector's polarity, you must flip a coin or depend on memory. Alternatively, you can return to the retailer, ask the clerk to show you the same unit you've left in pieces on the kitchen table. When his back is turned quickly pull out your jeweler's screwdriver, open the unit, and check to see just how the connector should be oriented. He won't mind...

7. *Crash past the booby traps and "one-way" assemblies.* Some electronic products are designed to require a special tool for disassembly or reassembly, and without that tool and the knowledge to use it, your intrusion will be visible and the warranty void. Even worse, some units use plastic "flex-tab" construction. To open the unit a tool must be passed through the port, or hole, to bend back a connecting tab before the connected parts will separate. Alternatively, you can use a screwdriver blade to pry between the parts, breaking those flex-tabs, and then when you've finished you can re-assemble the pieces using silver duct tape, a badge of technical skill and merit that can be recognized from some distance.

8. *Clean it in the dishwasher when you're done.* Actually, this will probably have no effect if you let the unit dry thoroughly before powering it. If it's still a bit wet, and you wanted a new scanner anyway, dry it in the microwave. That *will* do it.

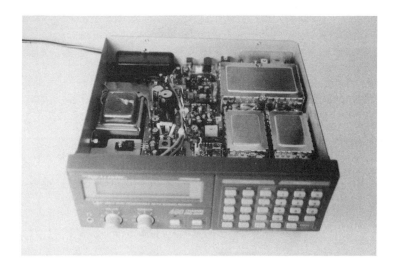

**The innards of the Radio Shack PRO-2006 – it opens with
an easy twist of an ordinary crowbar.**

Objectives and Procedures

1. INCREASE FREQUENCY COVERAGE

Not very many scanners offer continuous frequency coverage
between their lower and upper tuning limit. The majority
cover segments of the spectrum but leave gaps. The most
noteworthy such gap is, of course, in the band used by
cellular telephones and other Personal Communication
Systems (PCS). If one were to examine the frequency
planning of a synthesizer, the reason for the gaps is that it is
costly to cover the entire spectrum. For the techies, this won't
be enough. For the rest, it's too much. That's balance!

Circuitry such as that shown in *Figure 33* will produce gaps,
and there is no way to modify that circuitry to cover them.
For economic reasons, the designer downconverted the input
signal to two ranges using a single mixer.

A filter is used to separate the products of the mixer, but it's the nature of filters (physics) that the closer the frequencies the more difficult that task becomes. In the example shown, economics and physics force a gap in coverage near the mixer center frequency. The typical single-conversion radio is evidence that frequency gaps and price are inversely related.

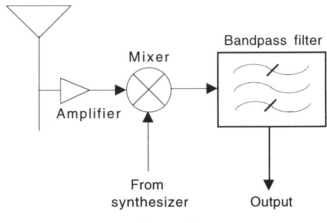

Figure 33

There's another way.

Suppose the design were to mix the signal twice, using overlapping bands to eliminate gap. Two LOs (local oscillators, or synthesized reference frequencies), two mixers, and two sets of filters are used. Each downconversion has its own gap, but because the LOs are different, the gaps appear at different points in the spectrum.

It's then the computer's job to (1) examine the commanded signal (from the sequence or from the operator's instruction), (2) select the downconversion circuitry that does *not* include a gap at the desired point, and (3) create the tuning instruction for the synthesizer to tune to the desired frequency. Many frequencies are covered by both downconversions; it's only at the gap points that serious computing is required.

No manufacturer wants to develop one product for the United States market, another for Europe, and a third for Japan and other countries in Asia. It's more profitable to make one design, but to include slight changes for each market. For instance, in some parts of the world it's illegal to listen to cellular telephone conversations; in fact, in the U.S. the government will no longer certify receivers that can tune cellular frequencies, but those frequencies are different in Japan, and different still in Europe. The "slight changes" might affect the RF/analog portion of the unit so it can no longer tune cellular in one market or another, or they might affect the instructions to the computer (usually on turn-on) as to which frequencies are tunable and which are not. In any case, it's the industry's common practice to prevent tuning of certain frequency bands with a reversible "fix." Reversal is often as simple as snipping one wire.

Several books have been published on the subject of scanner modifications, and they are included in the Bibliography. Perhaps the single most authoritative source of this information is Bill Cheek, whose books on scanner modifications are owned by a high percentage of scannists. In addition, Mr. Cheek publishes a newsletter (see Bibliography) that describes states of the scanning arts.

This chapter will discuss the frequency coverage modification of only a very few of the most common scanners. This information is provided merely to illustrate the level of difficulty of such changes.

Of course, neither the publisher nor the author take any responsibility for the result of actions you take to modify your scanner, and you are cautioned that not only can such modifications void your warranty, it may be a crime to modify a scanner to receive cellular telephone calls.

Tandy PRO-2006: With power OFF, disassemble the case. Directly behind the "3" key on the keypad is a diode marked D501, and near it is D503. Snip both from the board and remove them. Reassemble.

Tandy PRO-43: <u>With battery pack removed</u>, remove D-4 (restores cellular) and install it in the empty spots marked for D-3 (to restore 54-88 MHz).

Bearcat 200/205 XLT: <u>Remove the battery</u> and disassemble the case (two screws plus two more from the battery retaining plate). Open the case and carefully pull out the front panel. Make a sketch if memory might fail, and use a marker to indicate how the primary connector is oriented. Resistor R215 is just above the letters "DEN" on the main computing chip (the microprocessor) marked µC-1147 or UC1147. Snip both leads to that resistor and remove it. Reassemble.

Alinco DJ580T (a handheld transceiver, or HT, but with a scan function): <u>Remove the battery</u> and then remove two screws within the battery compartment and the plate they fasten, thus exposing a blue wire loop that is clearly longer than necessary and deliberately accessible. Snip the blue wire and reassemble. Re-initialize the unit (reset the CPU) in accordance with the operating instructions.

Bearcat 760: <u>With power disconnected</u> remove the four screws through the bottom cover. On the SANYO IC (LC3517BM-15), note the notch. With the notch UP, the pins are numbered from the top left corner, down the left and up the right sides. The bottom left pin is #16, and the bottom right is #17. Find pin 26. Cut the two traces leading to pin 26, solder together pins 26 and 27, then run a wire to connect pins 19, 20, and the two conducting traces that *went* to pin 26. Reassemble. Good luck.

Bearcat BC-950: This model is almost identical to the 760, above. First, <u>with power disconnected,</u> disassemble the case. Locate the microprocessor and clip pin 20. Reassemble. There are different manufacturing series that require different procedures. If you get an error message when entering a previously unavailable frequency, try the procedure for the 760.

2. ACCELERATE THE SCANNING RATE

The faster your scanner changes frequencies, the less you'll miss, right? So it's worth a lot to speed things up. There are two general ways to accomplish this.

First, remember that your scanner is a consumer product built to meet a price set by marketers for a given capability and set of functions. Production cost includes parts, labor, testing time, packaging, etc., but because labor is often the single most expensive part of the manufacturer's budget, it makes sense to have plenty of "margin" in the parts. That is, one selects parts that will be used at currents, voltages, and speeds well below their limits, thus reducing testing time.

For instance, a scanning circuit is driven/timed by a crystal that oscillates at some frequency. A 4 MHz crystal delivers four million clock pulses to the circuitry each second, and at each pulse the computer logic takes one more step. A competent production engineer will select logic that has plenty of margin, that can run at something close to twice the speed asked of it. He'll prefer parts that are specified by their manufacturer at perhaps 8 MHz. That criterion may increase parts cost somewhat, but it definitely reduces test time and rejects. It also creates an opportunity for the consumer.

To speed things up, you can replace the original logic timing crystal with a faster one. If you are comfortable *inside* a circuit, and have access to a frequency synthesizer instrument, use it as shown in *Figure 34* to determine the fastest frequency at which the devices that comprise the logic and microprocessor can be timed. If you wish to accelerate the system even further, you can seek the assistance of a competent technician with experience in digital circuitry. He may be able to look at the parts designators (often printed on the top surface of the chip) and identify a drop-in replacement from a faster logic family. Of course, if the circuit uses the same crystal to drive the synthesizer *and* the logic, this will not work without extensive modifications and a lot of courage.

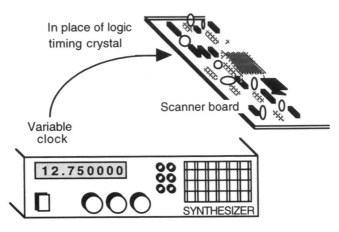

Figure 34

Once the acceleration modifications are complete, check the scanner at expected temperature extremes (get it hot and cold) to ensure that the logic still operates satisfactorily. In this case, the temperature extremes for a vehicle-mounted unit might be 0° F to +200° F, depending on your climate.

Most of today's products use surface mount technology (SMT), which means exceedingly small components soldered directly to the top of the circuit board (rather than wires passing through holes). Because it may be necessary to change resistors and capacitors as well as the crystal, if you take on this job you'd better have a fine flame on your propane torch. And if you don't have serious technical understanding, find someone who does.

3. IMPROVE FUNCTIONALITY

Scanner modification books and periodicals listed in the Bibliography contain an array of possible changes to your unit. They include ways to "fix" nearly everything.

For the more well-known intermediate and advanced products, there are many modifications. While you'll find plenty for the PRO-2006 and the Bearcat 760/950 and 200/205, there's nothing on the Yupiteru (at one extreme) or the Cobra (at the other). If you expect to get really "involved" in this hobby, stick with the brands that the experts thoroughly understand, and on which there is a substantial experience base. Here are just a few of the possible "functionality" changes to consider:

Disabling, or providing a switch to select, the "beep."

Extending battery life.

Adding a tone-call (remote squelch control) system.

Coupling the scanner to a computer to record frequency "hits," control tuning, etc.

Adding a spectrum analyzer function to permit viewing of signals in the spectrum.

Reducing step size, or tuning increments.

Increasing the number of memory locations (channels).

Improving squelch action.

Adding extra priority channels.

Adding an S-meter (to measure signal strength).

Improving control over the light (on a handheld).

And just to be safe, perhaps you should put a shield around the antenna to prevent cancer...

This book is obviously not intended as a step-by-step procedure to improve scanner action, but rather as a guide to (1) let the reader know what can be done, (2) what the process might entail, and (3) where to go for detailed modification procedures supported by photography. For further information, check the listings in the Bibliography. It's there.

And if you want the performance without dirtying your hands, there are companies and individuals out there that will happily do the job for you. Check the Bibliography and read the magazines, and if you don't see what you want, visit a local ham or scanner club and ask for help. You'll get plenty because everyone loves to be viewed as an expert.

The simple Velcro™ mod allows the scanner to be easily mounted anywhere, including upside down on the windshield (in those states where it's legal to have a scanner in a vehicle).

Using Your Scanner

So you finally picked a scanner, used it for a week, sold it to your brother-in-law, bought the one that has the functions you *really* wanted, and then had it modified by a guy at a swap meet. Now you're ready to justify all that money you spent.

The rest of this book was written to help you do just that. But before you step through the mirror, perhaps you should meet some of the folks from the tea party. There's more out there than a white rabbit.

THE FEDERAL COMMUNICATIONS COMMISSION 6

Uncle Charlie...

Purpose

The frequency spectrum is occupied by an enormous number of inhabitants ranging from Aircraft to Zoo security staff, and they all want to transmit at once. That community is governed by the Interdepartmental Radio Advisory Council (IRAC), which handles federal allocations, and the Federal Communications Commission (FCC), which allocates spectrum to all non-federal users, whether law enforcement, amateur, radiotelephone, broadcast, industrial, or zoo security. The FCC licenses television stations, assigning a specific channel (frequency) and permitted power output. It licenses amateur radio operators, or "hams," giving each a unique call sign that is a mandatory part of the ham's dialog on the air. Every aspect of radio operation is controlled/licensed/monitored by the FCC, and even a clock radio will be licensed because the circuitry of virtually all receivers will radiate a measurable amount of energy.

Computers are licensed by the FCC for the same reason, as are many other commonplace electronic devices in our environment. When it comes to emitted radio energy, the FCC is the judge, jury, motor vehicle department, and highway patrol. The FCC is responsible for the "quality" of the spectrum, and governs the community that inhabits it.

There is only so much spectrum. The band from near-zero to the upper limit of radio technology includes only so much room in it. It's meted out carefully by the FCC, and when a new splinter of the spectrum is allocated to some application, the result can be the creation of a new industry. Spectrum is precious, and the closer together signals can be, the more of them can fit in any portion of the spectrum. The electronic industry is constantly moving toward conservation of spectrum through use of more and better frequency synthesizers and crystal-controlled radios, since that packs more and more into the available space.

As entropy insists that our universe constantly moves toward chaos, the FCC requires order and discipline – and therefore conflicts with nature. Congress has given the FCC the power to stand up against Mother Nature (and the CB violator), and there *will* be order and discipline across the spectrum.

Policing the Spectrum

The FCC maintains (the wrong term, considering their budget) thirteen field monitoring stations across our nation. A typical station will include one or more buildings, plus extraordinary outside antennas, and will usually operate two or more mobile monitors in trucks. The station is filled with radios, of which many are scanners of one type or another.

Virtually all the monitoring equipment used by the FCC is available to the consumer, including antennas, radios, and computers, though some of the software might be hard or even impossible to find. When it comes to software, there's plenty on the market that even the FCC might find useful.

The most prominent antenna at a monitoring station is the circular array. It consists of a circle of towers, usually metal (sometimes wood, like power poles), strung with wire, with a diameter as large as 500'. Such an antenna can determine the direction from which a signal is received with an accuracy of only a few degrees.

Other antennas are used. Very high gain (thus directional) antennas permit both excellent sensitivity and high directivity, and the typical monitoring station will use many, each optimized for its own segment of the spectrum.

The building houses tape recorders, computers, and a variety of mechanisms to control the rotatable and switchable antennas around the station.

What do such stations look for? Their job is not to monitor live television broadcasts for the "seven deadly words," since they know that the first utterance will generate a deluge of phone calls from hundreds of self-righteous assistants. Whether working alone or in concert with other monitoring stations, they actively seek:

Pirate (i.e. unlicensed and unauthorized) radio, where hobbyists transmit music, commentary, and their own slant on the news. Political position is not as important to the FCC as is their potential interference with licensed stations.

Jammers, whether malicious or accidental.

Procedural violations by licensees. For example, hams who fail to identify themselves.

CB operators with linear amplifiers that transmit signals hundreds of times more powerful than allowed.

Systems and operators that transmit outside their licensed segment of the spectrum.

Nazi spies.

Penalties

It's serious business, because the FCC can levy fines and confiscate equipment. Some radio violations are punishable by imprisonment.

Some FCC penalties are quite significant. If a broadcaster seeks immortality by broadcasting a hoax (remember Orson Welles?) the fine is $25,000. Fake distress calls might cost up to $8,000 for each day they are transmitted.

Perhaps most significant is that the FCC has been given the right to assess the ability of the perpetrator to pay, and to set some fines accordingly. That is, if an "economically average" person were to add an illegal amplifier to his CB, the fine would be $5,000 per day. If the FCC decides that the violation is by someone to whom that figure would not be a sufficient disincentive, the fine can be elevated by up to 90%. Further, if the FCC determines (to its own satisfaction) that the misconduct was flagrant and deplorable, that justifies up to a 90% increase.

The FCC, as can be seen, uses fines as a powerful control mechanism.

Detectives of the Æther

Detecting an illegal transmitter mounted on a moving platform (aircraft, truck, boat, etc.) is difficult and depends upon a combination of skill, equipment, and luck. If an FCC monitoring vehicle happens to be near a truck that is transmitting with an illegal linear amplifier ("foot-warmer"), that's luck.

It's easy when the transmitter is stationary, as FCC monitoring stations can receive over very long distances and they work together using computers to integrate data from multiple sources. The combination of highly directional antennas plus sensitive receivers is so efficient that long-distance triangulation can be very effective.

If a transmission is detected by three or more monitoring stations, even if they are hundreds of miles apart, computers will correlate their data to determine the transmitter location with excellent accuracy. Once the general location is known, mobile monitors can usually track down a transmitter. Unfortunately, funding for those vehicle-mounted receivers is waning, but as long as they are available the persistent violator will eventually be caught.

As of late 1992, the FCC's monitoring stations were in:

Anchorage, AK	Powder Springs, GA
Ferndale, WA	Allegan, MI
Livermore CA	Vero Beach, FL
Douglas, AZ	Laurel, MD
Honolulu, HI	Belfast, ME
Kingsville, TX	Santa Isabel, San Juan, PR
Grand Island, NE	

Field offices of the FCC are in:

Seattle, WA	St. Paul, MN
San Francisco, CA	Detroit, MI
Long Beach, CA	Baltimore, MD
San Diego, CA	Philadelphia, PA
Miami, FL	Chicago, IL
Washington, DC	Kansas City, MO
Buffalo, NY	New Orleans, LA
Denver, CO	Tampa, FL
Dallas, TX	Boston, MA
Houston, TX	Norfolk, VA

You can call your local field office (the phone directory – under "U. S. Government") to complain about a neighbor's CB rig jamming your television set, or to complain about an offensive radio program. Experience has shown that FCC officers are professional and helpful, and rarely bust down doors. Yet they are the police force of the spectrum, with responsibility for the quality of our electronic environment.

That quality is diminishing, and will continue to do so regardless of how careful the FCC is, or how well it is funded, because the more transmitters there are, the higher the background noise level. The development of cellular and cordless phones, pager services, wireless security systems, and other generators of radio-frequency energy has corrupted the spectrum. It's like the background light of a city reducing the effectiveness of a nearby observatory. The higher the noise, the worse the performance.

As you will see, as a police force the FCC's "beat" is huge, yet it does a commendable job of regulating a critical aspect of our society, and that's despite a diminishing budget and aging equipment. Credit must be given to the staff, because it's easy to see that everything else is about worn out. As the reader will discover after a few days with a scanner, noise is a serious problem, and our government cannot afford to shrink the FCC to a level where it can no longer ensure the spectrum quality required by evolving wireless technology.

The risk to the spectrum is similar to the risk we accept by allowing certain industries to pollute our waterways. Eventually the medium, whether the æther or the river, will be so corrupted that it cannot be used without special equipment and expensive treatment. It's the job of the FCC to ensure that the spectrum remains clean and properly managed, like our woodlands, our waterways, and our other natural resources.

That takes funding, which has diminished steadily over the years though the requirement has increased, because the FCC is one of the least squeaky wheels in our government (their task is to *listen*). It's not a disaster, so we simply must make intelligent decisions, based on good information, and then accept the consequences of our actions. In this case, that end result will be a regrettable – *and expensive* – corruption of an important resource.

THE SPECTRUM 7

General Allocations

This chapter describes how much of the scannable portion of
the spectrum has been divided. That information is the key to
some of the text that follows, which will translate some of the
terms and expand the acronyms. Readers *are authorized* to
copy and even enlarge these pages for their own use.

The table omits many splinter/specialty users of the
spectrum, so don't depend entirely upon that information, as
new allocations are frequent and some are geographic,
varying from one part of the country to another. Though far
from perfect, the following table is a useful map of an
interesting world. All frequencies are in MHz unless
otherwise noted.

0 Hz-6 kHz	Extremely Low Frequency (ELF)
6-30 kHz	Very Low Frequency (VLF)
30 kHz-300 kHz	Low Frequency (LF)
300 kHz-3 MHz	Medium Frequency (MF)

Above 30 MHz, many bands are "channelized," whereby
specific channel spacing is assigned by the controlling
agency, established by the FCC, or agreed by convention
among the users.

In the table that follows, there is considerable overlap where
more than one user is licensed to use a specific channel or
band; in such cases, separation is often geographic. Where no
channel spacing is indicated, a variety of protocols may exist.
Also, there are many gaps where frequencies have not been

allocated to a permanent user, hence the usage is not published or is not available under Freedom of Information legislation. Finally, temporary allocations and abandoned licenses mean that many assignments will conflict with other entries in this book. The scannist's best friend is a *current* frequency list, augmented by a subscription to each of the important periodicals. This list is useful, but is intended to be no more than a guide.

Frequency band (MHz)	Steps (kHz)	Agency, or usage
30-50	10	Military VHF low band
30.01-30.55	20	Federal government
30.58-30.64	20	Industrial
30.66-30.82	40	Petroleum
30.86-31.02	40	Forestry
30.86-31.08	40	Motor carrier
31.10-31.14	20	Motor carrier
31.28-31.96	40	Industrial
33.02-33.10	40	Highway maintenance
33.18-33.28	20	Petroleum
33.44-33.68	40	Fire service
33.46-33.66	40	Fire service, mobile
33.70-33.98	20	Fire service, base/mobile
34.01-34.99	20	Federal
35.28-35.52	40	Industrial
35.74-35.86	40	Industrial
36.01-36.99	20	Federal
37.04-37.32	20	Municipal police
37.46-37.58	20	Power service

Frequency band (MHz)	Steps (kHz)	Agency, or usage
37.62-37.82	20	Power service
37.90-37.98	20	Highway maintenance
38.27-38.99	20	Federal
39.08-39.24	20	Municipal police
39.48-39.98	20	Municipal police
40.01-40.99	20	Federal
41.01-41.99	20	Federal
42.02-42.16	20	State police
42.18-42.30	20	State police
42.32-42.64	20	State police
42.66-42.70	20	State police mobile
42.80-42.94	20	State police
43.28-43.52	40	Industrial
43.70-43.84	20	Motor carrier (passenger)
43.86-43.94	20	Motor carrier (property)
43.96-44.34	20	Motor carrier (property)
44.36-44.44	20	Motor carrier (property)
44.48-44.60	20	Motor carrier
44.64-45.84	40	Forestry
44.78-44.90	40	State police mobile
45.08-45.64	40	Local government
45.46-45.70	40	Municipal police
45.68-45.84	40	Highway maintenance
45.90-46.02	40	Municipal police
46.06-46.20	20	Fire service

Frequency band (MHz)	Steps (kHz)	Agency, or usage
46.22-46.34	20	Fire service mobile
46.36-46.50	20	Fire service
46.52-46.58	20	Local government
47.02-47.40	20	Highway maintenance
47.44-47.68	40	Industrial
47.70-48.54	20	Power service
48.56-49.58	20	Forestry
48.56-49.50	20	Petroleum
49.52-49.58	40	Industrial
49.61-49.99	20	Federal government
50-54		6-meter HAM band
54-72	6 MHz	TV channels 2, 3, 4
72-76		VHF mid band
76-88	6 MHz	TV channels 5, 6
88-108	200	Broadcast FM
108-137		VHF aviation band
136-174		VHF high band
136-138		Satellite communication
138-144		Special military operations
144-148		2-meter HAM band
148-150		Special military operations
150.815-150.890	15	Automobile emergency
150.905-150.965	15	Automobile emergency
150.995-151.130	15	Highway maintenance
151.145-151.490	15	Forestry

Frequency band (MHz)	Steps (kHz)	Agency, or usage
151.520-151.595	15	Industrial
152.270-152.450	15	Taxicab radio
152.465-153.395	15	Forestry
152.870-153.035	15	Industrial
152.870-153.020	30	Motion picture
153.035-153.395	15	Petroleum
153.050-153.395	15	Manufacturers
154.130-154.250	15	Fire service
154.310-154.445	15	Fire service
155.520-155.700	15	Municipal police
156.045-156.075	15	Highway maintenance
156.105-156.135	15	Highway maintenance
156.165-156.240	15	Highway maintenance non-state
157.050-157.175	25	Marine radio emergency
157.200-157.400	25	Marine radiotelephone
157.470-157.515	15	Automobile emergency
157.530-157.710	15	Taxicab radio
158.280-158.325	15	Forest products
158.280-158.325	15	Manufacturers
158.280-158.355	15	Petroleum
158.985-159.075	15	Highway maintenance non-state
159.105-159.165	15	Highway maintenance non-state
159.225-159.465	15	Forestry
159.495-160.200	15	Motor carrier city to city
160.215-161.565	15	Railroads

Frequency band (MHz)	Steps (kHz)	Agency, or usage
161.640-161.760	30	Remote broadcast pickup
161.800-162.000	25	Marine radiotelephone
162.025-173.175	25	Federal government
173.225-173.735	50	Press relay
173.225-173.375	50	Motion picture
174-218	6 MHz	TV channels 7-13
220-222		Land mobile
222-225		1.25 meter HAM band
225-400	25	Military aviation
450.050-450.950	50	Remote broadcast pickup
451.025-451.275	25	Power service
451.175-451.275	50	Forest products
451.175-451.275	50	Forestry, petroleum, power
451.300-451.675	50	Telephone maintenance
451.375-451.675	50	Power service
451.375-451.475	50	Forest products
451.375-451.675	50	Manufacturers
451.375-451.525	50	Petroleum
451.525-451.570	25	Forest products
451.550-451.700	25	Petroleum
451.825-452.025	20	Industrial
452.325-452.475	50	Local government
452.525-452.600	25	Automobile emergency
452.775-452.875	50	Local government
452.900-452.950	25	Local government

Frequency band (MHz)	Steps (kHz)	Agency, or usage
453.050-453.950	50	Local gov't, police, fire, hwy
454.025-454.650	25	Mobile telephone UHF
454.675-454.975	25	Mobile telephone (aircraft)
455.050-455.950	50	Remote broadcast pickup
456.025-456.275	25	Power service
456.175-456.275	50	Forest, mfgrs, telephone, petrol
456.300-456.675	50	Telephone maintenance
456.375-456.675	50	Power service
456.375-456.475	50	Forest products
456.375-456.675	50	Manufacturers
456.375-456.525	50	Petroleum
456.525-456.750	25	Forest products
456.550-456.700	25	Petroleum
456.825-457.025	20	Special industrial
457.325-457.475	50	Railroad
457.525-457.600	25	Dockside longshoremen
457.775-457.875	50	Railroad
457.900-457.950	25	Railroad
458.050-458.950	50	Local gov't, fire, forest
458.050-458.950	15	Highway maintenance
459.025-459.650	25	Local telephone UHF
459.675-459.975	25	Local telephone UHF A/C
460.025-460.500	25	Municipal police
460.525-460.675	25	Fire service
460.650-461.000	25	Business airlines

Frequency band (MHz)	Steps (kHz)	Agency, or usage
462.200-462.525	25	Manufacturers
462.550-462.725	25	General mobile
465.650-466.000	25	Business airlines
466.026-467.175	25	Business
467.200-467.525	25	Manufacturers
467.550-467.725	25	General mobile
467.750-467.825	25	Dockside longshoremen
468.200-469.925	25	Business radio
470-512	6 MHz	UHF TV, channels 14-20
512-806	6 MHz	UHF TV, channels 21-69
806-821		Land mobile repeater inputs
824-846	30	Portable cellular phones
846-869		Land mobile repeater outputs
869-894	30	Cellular telephone
894-896		Aero mobile, public mobile
896-902		Land mobile, repeater inputs
902-928		Personal comm, 33cm HAM
928-935		Paging
935-941		Land mobile, repeater outputs
941-944		Government
944-950		Studio transmitter links
950-960		Fixed public, private microwave
960-1215		Government
1850-1990		Emerging personal comm
2110-2150		Emerging personal comm

A visual display of the spectrum can be inexpensively purchased from the Government Printing Office. It's a wall chart, poster size, and it displays reasonably current FCC allocations in great detail. The GPO also publishes a variety of references on the spectrum and its use, and publishes a catalog that is FREE for the asking: write to

> The Superintendent of Documents
> Government Printing Office
> Washington DC, 20402

Characterizing spectrum usage requires a bit of study. A spectrum analyzer will tune, detect, and display energy at every frequency at which it appears, but frequency alone is not enough to support scanner operation. To scan effectively, the hobbyist must know the following:

Frequency range (upper and lower limit). For instance, a band dedicated to a usage of interest might be described as "88.00 MHz to 108.00 MHz."

Channel steps, or the spacing between frequencies used within the band. A given usage might be described as covering its range in 10 kHz steps. Whatever that number is, the scanner should do the same, or a submultiple. When looking for *any* signal, the finest step size should be selected.

Modulation. Most scanners are capable of demodulating either Amplitude Modulation (AM) or Frequency Modulation (FM). FM can be either wideband (WFM) or narrowband (NFM) and they require different circuitry.

A more complete description of, for instance, the military UHF band might include the following: "225 MHz to 400 MHz in 25 kHz steps, using amplitude modulation." The data can be more explicit (modulation rates and other technical characteristics, etc.) but that is generally unnecessary to support scanning with conventional equipment available to the consumer.

Repeaters

A repeater... repeats. It receives radio signals and re-transmits them, usually on a new frequency, with more power than the original transmitter and/or a better line-of-sight to the receiver, and therefore with more range. The range is enhanced by geometry because repeaters are generally placed atop tall buildings, hills, and even mountains.

The repeater is often connected to the commercial telephone system, which allows the radio operator to use a portable radio to reach the repeater, and through it call home. Some particularly sophisticated systems allow calls to be made to the portable radio, which receives a transmission and recognizes a special signal embedded in it (modulation) that indicates that the transmission is intended specifically for it. It's no substitute for a cellular phone, but it works well enough for hams and some commercial radio users.

One of the conventions in popular frequency books is to add the letter "R" to frequencies known to be repeated, which indicates to the scannist that the signal is receivable from almost anywhere. Many VHF frequencies are repeated, particularly those depended on by industrial land mobile users, and mobile systems in the 440 MHz range and above are almost always repeated. Repeating isn't simple, though.

If both the receiver and transmitter were to operate at the *same* frequency there would be a closed loop of feedback as the receiver receives the transmitted signal that it just received, etc., and the system won't work any better than a government agency that reads its own reports and then reports on what it does.

To work properly, a repeater must receive at one frequency and then retransmit at another one; it's too bad there's no similarly simple solution for the government. There are conventions for repeaters in various bands, too, as shown in the following list:

30-50 MHz – the input frequency is about 0.5 to 2.0 MHz *higher* than the output frequency (i.e. 35.0 MHz base, 36.5 MHz mobile). Because of the ambiguity, one must scan to find the pair. Fortunately, there is very little repeating in this band because range is excellent, attributable to bending of the radio signal even over the horizon. Almost all domestic communication in this band is simplex.

144-148 MHz – in the lower part of this band, the input frequency is usually 0.6 MHz *lower* than the output frequency (i.e. 144.700 base, 144.100 mobile), reversed in the upper portion.

148-174 MHz – the input is usually 3.0 MHz *higher* than the output (i.e. 152.000 mobile, 155.000 base).

222-225 MHz – the input frequency is exactly 1.6 MHz *lower* than the output.

450-470 MHz – the input frequency is exactly 5 MHz *higher* than the output.

470-512 MHz – the input frequency is exactly 3 MIIz *higher* than the output.

806-896 MHz – the input frequency is exactly 45 MHz *lower* than the output.

935-941 MHz – the input frequency is exactly 39 MHz *lower* than the output.

To tune a communication that is being repeated, program both the input and the output frequency into appropriate channels. Obviously, one could enter a variety of pairs into a bank, and then use the scan button to follow a dialog, but it's easier to simply scan two frequencies per bank.

Repeaters work as range extenders. The scannist can view them as funnels through which flow concentrated streams of ham, dispatch, or mobile industrial communicators in the area, with your scanner waiting patiently at the narrow end.

Karl Marx
Communist Manifesto
(1848)

"The bourgeoisie, by the immensely facilitated means of communication, draws all, even the most barbarian, nations into civilization."

AMATEUR RADIO

Allocations

The amateur radio communication bands, assigned by the FCC and competently managed/monitored/policed by the hams themselves, are:

Range (MHz)	Band
1.800 - 2.000	160 meters
3.500 - 4.000	80 meters
7.000 - 7.300	40 meters
10.000 - 10.150	30 meters
14.000 - 14.350	20 meters
18.068 - 18.168	17 meters
21.000 - 21.450	15 meters
24.890 - 24.990	12 meters
28.000 - 29.700	10 meters
50.000 - 54.000	6 meters
144.0 - 148.0	2 meters
222.0 - 225.0	1.25 meters
420.0 - 450.0	70 centimeters
902.0 - 928.0	33 centimeters
1240 - 1300	23 centimeters

As can be seen, our government has been generous in allocating spectrum to the amateur radio hobbyist. With spectrum among the most valuable and critical of resources, and considering that allocation of even a narrow band by the FCC often creates an industry (as happened with pagers, for instance), it is interesting to note that the amateur services occupy a high percentage of the available frequencies, exceeded by broadcast radio and television (the largest users), the government (including military), and business. The list above identifies the most scannable of the amateur bands. Much more explicit information, including the classes of operators and the types of communication authorized, can be found in several of the American Radio Relay League (ARRL) books listed in the Bibliography.

The Amateurs

More than a half-million Americans are Amateur Radio Operators, licensed by the FCC and policed mostly by each other. In these days of chaotic sociological stresses it is a pleasure to enter the world of generally ethical, honest, and straightforward HAMs. Amateur radio enthusiasts pursue their hobby for their own entertainment, to support private communication needs, and to serve their community. That service has long been a resource to emergency agencies, the military, and law enforcement, not only in North America but in many other countries as well.

By international treaty, the amateur radio service is authorized in virtually every country on earth, and monitoring, licensing, and standards are reasonably coherent from one country to another. Since radio signals ignore national boundaries the amateur radio community is truly global. Around the world, more than 1,500,000 licensed operators share news, gossip, technical tips, and disaster relief information. Even the Soviet Union of the 1980's had more than 20,000 operators, and the government actually considered amateur radio as much a sport as chess!

The frequency allocation table shows most of the specific segments of the spectrum dedicated to amateur radio, and how they relate to neighboring users. It's easy to see why the amateur bands are so jealously guarded by its occupants, who support some of the most vocal lobbyists in Washington.

In this country, a ham license can be acquired by anyone who passes the written and practical tests, administered by other hams authorized to do so by the FCC's licensing facility in Gettysburg, PA. The tests vary for each license class, include technology, frequency usage, and all but one require passing a Morse Code test as well. Age, race, social status, financial means, disabilities, religion, all have no bearing on the issue, though it helps to be tall if you're planning on an outdoor antenna. When you pass the test you get a license, and along with it your unique call sign.

The amateur bands are public, non-business, and open to anyone who cares to listen. No one who uses these bands expects privacy, so there is no ethical question about listening. In fact, there is a whole class of short wave listeners (SWL) whose hobby it is to tune, and scan, the amateur frequencies. The scannist won't want to miss the amateur bands, and to understand who and what is overheard requires a bit of inside information, because the call sign is more than an identifier. A U. S. call sign that begins with A, K, N, or W is followed by a number that indicates the geographic area in which that call sign was issued (and, therefore, the likely source of the transmission). Check the Bibliography for ARRL references, which include geographic allocation of call signs, and much more.

The Hobby

The term "amateur radio" might conjure up the image of a "ham shack," wherein a table is covered by radios with huge dials, the wall is covered with postcards proving signals received, and the ham himself (they're still almost all male), wearing earphones, hovers over his equipment making microscopic adjustments as he strains to hear a faint signal from around the world. Today's amateur is likelier to be wearing his "ham shack" on a belt clip. Called a handheld transceiver (HT), his radio looks much like a small cellular telephone, but it has many more buttons and controls than any cellular, and does much more, also.

HTs generally operate in the 2-meter band, 144 MHz to 148 MHz. That's a popular portion of the spectrum, and many repeaters are operated by ham groups and commercial services to permit wide range in this and other bands.

The second most popular band for today's hams is the 70 centimeter band, or 420 MHz to 450 MHz. Nearly all operations in this band use repeaters and autopatch (an automatic connection to the telephone system), and most ham experts use portables that work in this range.

Ham phone calls to home are clear and easily understood, but ham-to-ham communication often uses terms that may not be familiar to the neophyte listener, so the Glossary includes some common terminology. The best of the amateur radio books are published by the ARRL. Their address, and a brief list of books they offer, appear in the Bibliography.

For those who enjoy listening to hams in action, one of the most useful ARRL books is the $6 *Repeater Directory*. It is a reference manual that lists virtually all authorized ham repeaters in the country, with information on how to contact local frequency coordinators (regional gurus whose task it is to control the airwaves to avoid interference among hams), and – of great interest to prospective hams – a list of all ARRL clubs, with meeting times and locations. It lists more than 19,000 repeaters nationwide, so it is difficult to find any urban area that is not served by at least one or two of them. What good are they?

A repeater was defined as a radio that listens to a weak signal from a handheld transceiver and re-transmits it, usually from a high vantage point, so it can be received throughout the area. Ham transmissions are therefore "amplified" by repeaters. And since the repeater repeats a multitude of weak transmissions from a broad area, it creates an opportunity for the scanner, which can be programmed to rapidly switch between the input and the transmit frequencies of a repeater.

For hams, and a few authorized others, the repeater usually adds an "autopatch" function, which permits the ham to enter the phone system and use his H/T as a conventional telephone. These radio conversations are usually casual and friendly, and deal with almost any matter except business (strictly forbidden). Every ham must introduce his/her call sign into the conversation at the end of each transmission or series of transmissions, or at least every ten minutes. That call sign is a true identifier, and it is therefore possible for even a casual listener to learn identities and even likely addresses of those on the air by contacting local clubs or using a more formal process involving the FCC.

Hams transmit and receive more than voice and Morse code (CW). Teletype, video images, computerized packages of data (packet), and satellite communications are all within the scope of the advanced amateur, and there's plenty of specialized equipment on the open market that will make these transmissions available to the casual scanner enthusiast.

Public Service

In addition to providing communication during outright disasters, hams regularly provide such public services as the reporting of road and weather conditions, and a properly equipped mobile ham will summon help for motorists in distress. Two national organizations are: Radio Amateur Civil Emergency Service (RACES), and Amateur Radio Emergency Service (ARES), and both are important resources in emergencies. A member might sit in a "radio shack" (that's where the name came from!) monitoring almost everything, or might drive four-wheel-drive vehicles festooned with antennas, tools and fire extinguishers. The dedicated member is hard to miss on the road, and his radio transmissions during flood, earthquake, or other civil disaster are easily identified by a unique professionalism. Members may invest thousands of dollars to equip their "shack," and many hours in training, but they are today's most highly paid communications experts though they earn *only* the good will and thanks of those who benefit from their effort.

Monitoring Ham Radio

There's lots to be heard by monitoring hams with a scanner, and many scanners will easily tune many ham bands. Focus on this portion of the spectrum can become infectious in two ways, especially if there are several repeaters and ham clubs in your community. First, there's lots to hear and therefore listening can soak up many hours at the expense of other activities. Second, listening to hams is the first step to becoming one, and that will use up even more time and financial resources. Frankly, as literally hundreds of thousands of hobbyists will confirm, it's time and money well spent.

This book discusses "gray markets," defined as those wherein one can make/sell/buy and own something that is usually illegal, immoral or unethical to actually use. That definition absolutely does not apply, in any way, to the amateur radio community or the legitimate businesses that support it. In fact, hams police their own population very effectively, and are so communicative that questionable businesses can become well known very quickly, and shunned effectively.

By the rules of the FCC, ham transmissions are freely accessible by anyone with the right equipment. There is nothing even slightly questionable about the fascinating hobby of listening to hams in action, and the hams welcome the company.

MILITARY RADIO

9

Standardization

The United States military has established standards that have been generally adopted by many nations, friend and foe. That may be because much of the world's military communication equipment is produced in this country, but in any case almost all military aircraft use 225 MHz to 400 MHz in 25 kHz steps, with amplitude modulation, and most tactical battlefield voice communication is in the 30-50 MHz range. Some such transmissions are scrambled, but most are in the clear and are both informative and enjoyable. There were once bans on monitoring the military, but federal law has been liberalized and local laws on the subject are rare.

Spectrum Usage

In our great nation, military requirements have priority over everything else in the spectrum. Except entertainment, of course, and as CNN proved during Desert Storm, technology has blurred the distinction between war and boxing. Military frequency allocations are divided into functional categories:

MILSTAR: multi-service satellite communication	43 GHz
Other satellite communication	Ku-band
Satellite datacom	X-band, 1kHz steps
Satellite position location (GPS)	L-band, fixed frequency
Unpiloted airborne vehicles (UAV)	mostly L-band
Military air	225-400 MHz
Ship-to-ship (close-in)	UHF, VHF
Special military systems	148-150 MHz
Base security	VHF, low-UHF
Battlefield tactical comm	30-76 MHz in 50 kHz steps
	30-88 MHz in 25 kHz steps
Ship-to-ship/shore (long range)	2-30 MHz (HF)
Submarine communication	ELF, <<100 kHz

Opportunities

Within these allocations, among the most interesting is the 225-400 MHz air-to-air/ground traffic (a term that means "radio transmissions") overheard near military aviation training areas. Near Pensacola FL, for instance, one can hear the warm, helpful, and supportive voices of Navy flight instructors as they gently persuade their airborne charges to avoid mid-air collisions. Near the Naval Air Station, Fallon (Reno, NV) one can listen to equally encouraging instructors from the Top Gun school as they admit that it was only the students' bad luck that produced six straight "down-in-flames" computer victories.

The Marine Corps' Camp Lejeune, near Jacksonville NC, is another fine opportunity to listen to military leadership in action. Near 40 MHz, one can hear new second lieutenants valiantly lead their troops through simulated firefights, demonstrating their intellectual and educational superiority as they rise above the advice of their platoon sergeants. The simulated enemy can also be heard radioing for emergency helicopter diaper resupply as they capture one second lieutenant after another.

Fort Bragg's JFK Special Warfare Center, near Fayetteville NC, trains the Army's Green Berets, and gives the scannist frequent opportunities to listen to primary parachute training, which is often a real scream.

Norfolk VA, Long Beach CA, and San Diego CA are the Navy's busiest ports. It's easy to find frequent ship-to-shore communications, and with a harbor and mooring chart as a guide (available from most nautical supply shops) one can follow our brave sailors as they return from months at sea and give up command of their ships to the Harbor Pilots, who navigate the vessels to the appropriate mooring points. If the scanner is then tuned to the law enforcement frequencies, some of those sailors can be tracked from bar to "social club" to jail. It's nice to confirm that some things never change.

Ships are at their slowest when being pushed and tugged to the dock, so it's rarely interesting to listen to the occasional radio commands from the Pilot to the tugboats. On the other hand, it can be exciting to monitor the return of an aircraft carrier after a six-month deployment if one's husband is aboard. Or, in these changing times, if one's husband's boyfriend is aboard... *It's not just a job – it's a date!*

Military convoys communicate within the 30-50 MHz band, as do many base security functions, Post Exchange operations, and other non-tactical stations. The majority of ground military communication is in this range. In what the military calls "FM radio," specific bands of interest (all in MHz) are:

29.8 - 30.56	32.0 - 33.0	34.0 - 35.02
35.98 - 37.02	38.0 - 39.0	40.0 - 42.02
46.6 - 47.0	49.6 - 50.0	

Military aircraft equipped to carry VIPs, including politicians, usually have an air-to-ground radiotelephone capability. These include modified commercial airline hardware, such as the Douglas DC-9, MD-80, and similar conversions. They also include the venerable prop-driven Martin Marietta C-130 for lesser dignitaries, or for political inspections of areas where the airfield is too small for jets.

Frequencies used by military radiotelephones include either long-range HF, or UHF when in line-of-sight of a ground station. VIP air-to-ground radiotelephone operations are often conducted as follows:

Ground station transmissions .... 407.85 MHz

Aircraft transmissions .... 415.7 MHz

Long range (all in kHz) 6715, 6730, 6756, 6761, 6817, 9120, 11035, 11180, 11055, 13752, 15048, 16090, 18027

These frequencies are accessible only
to equipment that supports HF SSB.

VIPs

Identifying the person placing the call from a military aircraft is difficult, but there is a code system that will help identify his status. This was originated so that an arriving aircraft could warn the airfield to prepare sideboys and honors for the senior officer about to land.

Code 7   Colonel (Army, Air Force, Marine), Captain (Navy)

Code 6   One-star General, Commodore (Navy wartime rank)

Code 5   Two-star General, or Rear Admiral

Code 4   Three-star General, or Vice Admiral

Code 3   Four-star General, or Admiral

Code 2   Five-star General, or Fleet Admiral

Code 1   Michael Jackson/Jordan, or the President

When monitoring Navy harbor operations, the term "Ahoy" asks "Who's aboard?" The answer can range from the name of the ship or organization commanded by that passenger, to "Aye-aye," which translates to "no one but us peons."

A transmission that discusses a Code 7 identified as "Ranger" refers to the Navy Captain currently commanding the *USS Ranger*, though the captain of that particular ship will be wearing mothballs instead of gold braid.

Here are some other useful codes:

Air Force 1	The President's aircraft, and he's aboard
SAM-27000	The same aircraft, but he's *not* aboard (or this time he inhaled)
Air Force 2	The VP's aircraft, and he's aboard
SAM-21682	The same aircraft, without him
SAM-01	VIP flight carrying foreign head of state
SAM-8xxxx	VIP flight carrying a cabinet-level politician

Most air-to-ground radiotelephones are unscrambled. Since VIPs usually travel with a press retinue that uses radio communication to file stories, these frequencies can make for interesting listening. Remember, the users are a combination of military and political personalities, so this communication is for entertainment only and should never be taken seriously.

Air Shows

A military air show creates a fine opportunity for scannists. The participants rarely stop talking because the mandatory precision demands very close coordination in time and space, hence much voice communication. Air shows employ a ground controller, who observes from a vantage point that remains constant with respect to the flight pattern regardless of location. His comments are based upon experience, and can enhance precision and safety, and he provides the scannist with a stream of commentary that's more interesting than the booming voice from the PA system. A picnic near the airfield during rehearsals (normally 1-3 days prior to the show, and done to identify key landmarks for timing maneuvers) will easily find the frequencies to be used.

As a guide, the following table lists some frequencies used by the Navy's Blue Angels and the Air Force's Thunderbirds (now wouldn't *that* be a helluva dogfight?).

T-BIRDS	BLUE ANGELS	
295.7	281.6	302.15
273.5	275.35	263.35
283.5	307.7	
322.3	302.1	
322.6	345.9	
382.9	241.4	
394.0	245.9	
Ground support	Ground support	
413.100, 413.075	142.0, 143.0, 141.56	

This is a partial list of potentially interesting frequencies that are used nationwide by the military:

Guard Channel, for military aircraft in distress 243.0

Strategic Air Command 311.0, 321.0

Tactical Air Command 381.3, 320.0

USAF National Common 251.9

USAF Refueling 358.2, 361.7, 366.3,
370.4, 372.3, 396.2
(and many more)

Whether the listener works for a foreign intelligence service or is just blackmailing service wives, the military is a wonderful source of interesting communication. The Bibliography lists sources for frequency lists that would have been top secret a few years ago, but are open to the public today.

...sed quis custodiet ipsos Custodes? [1]

Juvenal

Basics

Government functions depend upon communication, and the IRAC (Interdepartmental Radio Advisory Council) is the primary federal agency that controls the spectrum in which wireless communication occurs. The IRAC assigns frequencies for government usage, and blocks go to the FCC for non-government allocations. Through hearings, and pro-active solicitations of opinion from lobby groups and experts (usually representatives of companies whose products and services use spectrum), both agencies "listen" to opinions and argument, and are generally considered responsive, fair, and competent. Nevertheless, usage conflict supports an army of attorneys specializing in FCC petitions and spectrum negotiation. In sum, the government is the largest non-entertainment user of our generally well-managed spectrum.

It is easy and legal to hear government radio communication, though one must listen long and carefully to identify the transmitting agency, and to understand the dialog. Frequency lists help, and most are published in "both directions." That is, they list the frequencies assigned to the user, and the users assigned to the frequencies, and one can look up the relationship between users and frequencies in either direction.

Government Frequency Allocations

The following table describes frequency usage (in MHz) by a few non-military federal agencies in California, though many frequencies are used nationwide by the agencies indicated.

[1] Who monitors the National Security Agency?

It is far from complete, and is provided as a limited guide and as an indication of the scope of government radio communication. Frequency lists and books are far more detailed.

Agriculture Department Special Agents
    168.6, 168.675, and 168.7
Alcohol, Tobacco, Firearms
    168.00
Civil Air Patrol
    148.15, 149.925
Department of Energy
    148.470, 149.145, 149.745
Drug Enforcement Agency, nationwide
    171.45, 171.65, 172.00, 172.100
Federal Aviation Agency (FAA)
    Nationwide repeater
    172.825
    Airport security nationwide
    165.5, 165.6625, 166.175, 172.15
Federal Bureau of Investigation
    120.425 (air), 164.8625, 165.5625, 412.425
Federal Emergency Management Agency
    142.35, 142.375, 142.425
Federal Executive Board Emergency Net
    170.2
Federal Protection Service
    417.2
Fish and Wildlife Nationwide Primary
    34.83
Forest Service
    166.675, 168.625, 170.0
General Services Administration
    164.275
Geological Survey (coastal)
    164.8
Health and Human Services Department
    413.425
Housing and Urban Development (HUD)
    165.6625

Interior Department Fire Cache
    166.725, 166.775
Interior Department Bureau of Land Management
    167.1, 168.25, 168.4
Interior Department Common Air Safety
    172.6
International Trade Commission
    410.725, 416.975
Interstate Commerce Commission
    409.2
Labor Department
    406.2
Maritime Administration
    166.15, 169.075
NASA's JPL/Pasadena
    163.0, 163.1, 410.0
National Highway Traffic Safety
    40.26, 40.97
National Oceanic and Atmospheric Administration
    162.425, 162.45
National Ocean Service
    162.10
National Transportation Safety Board
    165.75, 166.175
Nuclear Regulatory Commission nationwide
    165.6625
Park Service
    172.575
Postal Service
    162.225, 169.85
Prison Bureau Tactical, Emergency, Security
    170.65, 170.875, 170.925
Reclamation Bureau
    164.575, 164.725
Secret Service
    165-166 MHz
State Department Nationwide Nets
    407.6, 408.6

Treasury Department Common
   166.4625
Veterans Hospitals
   168.525, 168.575, 171.3875
War Department - Ravenhill 2-3014, ask for Mr. Stimson.

PLEASE WRITE TO THE PUBLISHER WHEN YOU FIND DISCREPANCIES BETWEEN THE FREQUENCIES LISTED HERE AND THOSE DESCRIBED IN OTHER SOURCES. SOME AGENCIES GET TEMPORARY ALLOCATIONS, MOVE AROUND, SHARE CHANNELS, OR OVERLAP OTHERS, AND STILL MORE DISAPPEAR. IN ANY CASE, THESE FREQUENCIES ARE OFFERED AS AN ENTRY POINT, AND IF YOU FIND ERRORS, THAT MIGHT PROMPT YOU TO BUY A PROPER FREQUENCY LIST FOR YOUR AREA...

Our Government – Business As Usual

Listening to government radio is interesting and useful only if the transmitting agency itself is doing something interesting and useful, which is relatively rare. The truth is that our government is not very boring because it's humorous when making mistakes, and they're both frequent and audible to the hobbyist with a scanner.

On a more serious note, it's useful to correlate listening to the government with the news services and law enforcement frequencies. When you hear of an event that might justify the effort, switching from news to law enforcement to the appropriate government agency can help develop a complete picture of what's happening.

For instance, if you hear a rumor about a reactor meltdown, switch to 165.6625 MHz. That frequency is *shared* between the Nuclear Regulatory Commission and Airport Security Nationwide, so if the Nuclear inspectors start calling ahead to bypass airport security, it's time for you to hire a Piper Cub and scoot.

CITIZENS BAND

11

How to Tune Citizens Band... and *Why*

There are several ways to tune CB. First, most shortwave receivers can reach that frequency band. All "continuous tuning" scanners (the R1, AOR 1000/2500/3000, Yupiteru, etc.) do also. Perhaps the lowest-cost way to reach the lowest-class communication system is to simply visit a swap meet and buy a CB rig. At the typical price of $10-20, you can buy two of the same brand and have some hope that at least one will work once the dried beer and peanut shells are cleaned out of the works.

And why do it? Because it's fun. It's a bit like that magazine you found in the seat pocket of the airplane, left behind in a taxi, or tucked behind the drawer in that motel room bureau (and you always check, right?).

CB is your chance to practice your holier-than-thou attitude. On the other hand, it's also your chance to learn something about how the system *really* works.

An Anomaly

Within the river of regulated, monitored, and well-managed communications is this truly strange island, with a language of its own and a Utopian population completely devoid of politicians, CPAs, attorneys, and IRS agents. It's full of pickup trucks, 18-wheelers, and good buddies, all with one thing in common: their enemy is Smokey the Bear. Citizen's Band Radio (CB) has become a subculture with its own rituals and lingo, and if you don't speak it you'll have to make a first-hand discovery of the perils just over that hill. Yes, there are published rules and regulations regarding CB, but one of the more interesting aspects of this subculture is that the rules are read mostly by those who write them, and by the occasional attorney hired to defend one of the few rule-breakers who are caught. Among the hundreds of thousands of CB users, there are more rear window rifle racks than people who know the rules governing their pastime.

Until about ten years ago, a license was required before a new CB'er could transmit for the first time. When the license application was filed, the applicant guaranteed that he had either bought, ordered from the Federal Government Printing Office or from the FCC, or otherwise acquired a true copy of Part 95 of the FCC's rules governing CB. Perhaps fewer than ten percent of the purchasers of CB equipment actually got the regulations, and fewer than ten percent of that group actually read them, and fewer than ten percent of that group actually applied for a license... and those few were probably writing books of their own. The attitude of the CB community is an indicator of the nature of CB.

Since a large percentage of CB radios changed hands at garage sales, and many cars and trucks were sold with the equipment still installed, a lot of CB users were unlicensed. Many CB'ers who proudly scoffed at the silly regulation that they own the regulation would be surprised to learn that since 1977 every CB sold had a copy of the regulations somewhere in the package – often it's that piece of tightly folded paper that kept the microphone from rubbing against the metal chassis. That document is probably read about as often as the instructions that come with a pocket knife. But the document, and the laws it revealed, exist and deserve at least honorable mention...

Legislation

But there *are* rules, and there is even an agency out there that tries to enforce them: the FCC, whose Rules and Regulations Part 95 constitute the authority and governing limits of the Citizens Band.

Title III of the Communications Act of 1934, as amended, allows the FCC to regulate radio transmissions and issue licenses. In concept, that Act applies to the Citizen's Band. The purpose of private short-distance radiocommunication is to support the business or personal communication requirements of citizens, and Part 95 establishes the rules governing those activities. Those rules also provide guidelines for the type acceptance of manufactured radios. Part 95 takes approximately twenty-five pages of fine print, and some believe the only time it is read is when the authoring staff checks to see whether changes are needed.

There were once four CB classes, and their evolution defines the ability of the FCC to adapt to changing times, technology, and user needs. The original rules described CB classes A, B, C, and D, but "B" was discontinued in 1971, and "A" has become the General Mobile Radio Service in the 462 and 467 MHz range, limited to 48 watts of output power and devoted to non-cellular radiotelephones and other "citizen" usage. Class C is intended for garage door openers, radio controlled model boats/cars/aircraft, and other remote control systems. What we call "Citizen's Band" today is therefore limited to Class D, created in 1958 when the FCC allotted 22 channels from what was previously the 11-Meter amateur band. Today, CB has been increased to 40 channels.

For the first few decades, when CB required a license, few seem to care who bought, ordered, or read, the regulations, or who paid their $4 fee for a license. That fee, by the way, was dropped because in 1976 some deep-thinking bureaucrat actually decided that COST was the reason people weren't submitting applications for official licenses! The forms changed frequently, at least seven or eight different forms existed, and some had the wrong mailing address for the FCC as well. When applications were returned marked "addressee unknown," the word got out, and no wonder the consumer became skeptical of the licensing procedure.

As of 1977, the FCC's CB Licensing Facility had a slow-moving backlog of over 250,000 applications. That backlog was eliminated by legislation that automatically grants a license to any U. S. citizen, without fee or examination. The end of the CB "license" was taken by some to be permission to operate without controls or regulation, or inhibition. In fact, the FCC *does* have authority over this band, but with resources limited by the budget the focus is elsewhere.

FCC's Frustration

The FCC's primary concern is the quality of the spectrum, and it's just about given up on the Citizen's Band. Budget limitations make it difficult to chase violators down the highway, which is where the majority of transgressions occur. On the other hand, the problems are restricted to a relatively narrow portion of the spectrum (though some CB'ers operate outside the assigned band), so wisdom dictates that the FCC's resources be expended on more attainable goals than cleaning up language on Citizens Band.

It's the most anarchistic and unregulated radio band in the United States, and also – to the scannist – among the most entertaining. One can hear hookers making arrangements, monitor the anti-Smokey program, count the hitchhikers on Highway 101, and listen to the meetings planned for the next Lenny's down the road (proof that advertising works – only ACLU attorneys use the name "Denny's" any longer).

Protection of – and respect for – Channel 9, the designated emergency channel, is perhaps the only way in which CB users consistently comply with the law. Channel 9 is reserved for true emergencies, and to coordinate REACT operations. Some of the most scofflaw CB'ers would skin alive anyone who decided for some distorted reason to corrupt that channel.

But how does the FCC zero-in on a truck buried in a stream of traffic, all moving at 65 miles per hour? It rarely can, but sometimes it finds a violator, confiscates his foot warmer, revokes his "license," and might even levy a fine.

Listening In

The FCC has essentially abandoned attempts to manage the Citizens' Band, and only the most egregious offenses deserve allocation from the FCC's limited (and shrinking) resources. If you easily become self-righteously indignant, listening to CB is not for you because you'll fume while waiting for the "authorities" to catch the offenders who prance across the CB spectrum. On the other hand, if you're objective and tolerant, CB can be fascinating. Next to the law enforcement channels, it's one of the most heavily used bands – especially near major highways. In a very real sense (common-law rules, language, idiosyncratic population) it's a culture, and even if you do not admire some of the behavior and language, CB is always at least interesting. Here's where to tune...

CB Channel Assignments

CH#	Freq	CH#	Freq
1	26.965	21	27.215
2	26.975	22	27.225
3	26.985	23***	27.235
4	27.005	24	27.245
5	27.015	25	27.255
6	27.025	26	27.265
7	27.035	27	27.275
8	27.055	28	27.285
9*	27.065	29	27.295
10	27.075	30	27.305
11	27.085	31	27.315
12	27.105	32	27.325
13	27.115	33	27.335
14	27.125	34	27.345
15	27.135	35	27.355
16	27.155	36	27.365
17	27.165	37	27.375
18	27.175	38	27.385
19**	27.185	39	27.395
20	27.205	40	27.405

*Channel 9 is the national emergency channel, used solely for emergencies (which to some may include running out of beer).

**Channel 19 is "Highway Primary," a good place to start by one seeking information or communication, or wanting to report sighting of a Smokey or a traffic hazard.

***Channel 23, or 27.235 MHz, is also used by Class C operators. Those who read the above paragraph know that Class C is reserved for such functions as remote control of model aircraft, and garage door openers. The FCC must have had good reasons to put those functions and the good buddies on the same frequency, but perhaps those old tales regarding first-generation (non-digital) garage door openers might really have some basis in physics.

Handles

FCC once assigned call signs to licensees, and there was a legal requirement to use that identity on the air. Today, one can scan the Citizen's Band for days without ever hearing anything but "handles." A "handle" is a self-assigned identity, often picked to highlight some physical, personality, behavioral, or professional/trade attribute. "Shorty" and "Slim" are easy to understand, but why would a rational human proudly announce to his peers that hereafter he is to be referred to as "Loser," "Malter Witty," "Wohn Jayne," "Scuzbucket," "2-4-Sex," "Nosepicker," "Grim Raper," or "Deadly Ugly?"

Collections of handles have been published, and there's little on the airwaves quite as funny – and sometimes as curious – as the handles people voluntarily assign themselves. Remember, it is the very nature of CB that grants anonymity to anyone who desires it.

Freud might have learned a lot by interpreting handles.

Language

CB lingo is an evolving language, and since it's used by a rapidly moving and relatively small population with little social inertia, it can evolve quite rapidly. New terms arise spontaneously and frequently, but it's considered gauche to ask on the air what, for instance, "riceburning superscooter" means. One can ask face-to-face, however, so the next time you see a bar with a bunch of modified motorcycles parked in front, go on in and ask the most heavily tattooed person you can find (perhaps a man) for a translation. He'll be glad to walk you through the subtle differences between a %#!!%& Ninja bike and a red, white, and blue Harley.

If strong language offends, stay away from the Citizen's Band. The FCC long ago gave up trying to catch users who fill the air with foul language. On the other hand, it's often funny, and sometimes can provide insights to human behavior that are simply unavailable from other sources. On CB, much meaning is conveyed by inflection that is impossible to describe using text alone, so even a perfect glossary is not fully sufficient. Understanding takes time...It's impossible to understand CB communications without a glossary or dictionary, and since the CB lingo is absolutely unique to the 27 MHz band it doesn't make sense to lump that translation dictionary with the rest of this book's glossary, so it's presented separately.

CB Glossary

All the good numbers – Good luck and best wishes!

Base – Base station, home or office installation

Bear – Any law enforcement officer

Bear bite – Traffic ticket

Bear Cave – Police station, highway patrol barracks, etc.

Bear-in-the-air – Airborne Smokey

CB Glossary, continued...

Beat the bushes – The first vehicle in a fast line of communicating vehicles, tasked to smoke out Smokey

Beaver – Female

Breaker – Anyone wishing to interrupt a conversation or make a transmission

Camera – Traffic/speed radar

Chicken coop – Truck scales

Clean – No known Smokeys in the area

C'mon – Please reply

Comeback – As in WWII cockpit talk: "Over," meaning "your turn to talk"

Copy – Statement = "I understand." Rising inflection = "Do you understand?"

Dirty side – East coast

Double nickel – 55 miles per hour

Down and gone – Off the air

Ears – CB rig, as in "Does he have his ears on?"

Eyeball – Visual contact

Flip flop – Return trip (truckers)

Floptop – Convertible, as "that beaver's in a floptop!"

Foot warmer – Linear amplifier to increase CB power and range (illegally)

Four – Abbreviated "10-4," meaning "I understand"

Flying a kite – Antenna higher than FCC legal limits

Good buddy – Another CB user, trucker, pal, redneck, etc.

Green stamp – Fee on a toll road, fines

Hammer down – Cruising faster than "double-nickel"

CB Glossary, continued...

Handle – CB nickname, preserves anonymity

Heater – Linear amplifier

Linear – An external amplifier to boost CB output power beyond legal limits

Local yokel – City or town law enforcement, as opposed to highway patrol

LSB – Lower sideband

Lurking – Passively listening to CB

Mercy! – An expletive, an audible exclamation point, shorthand empathy

Negatory – No

Off the map – Using a modified CB to reach illegal frequencies

On the side – Listening but not transmitting on that channel, lurking

Over your shoulder – Behind you, as in "Smokey's over your shoulder!"

Pedal to the metal – Driving far past double nickel

Picture taker – Police with radar

Plain wrapper – Unmarked police/patrol car

Punch through – Transmit effectively despite interference, usually with a linear amp

Ratchet-jaw – A good buddy who rarely stops talking

Reefer – Refrigerated truck

Rig – CB radio installation, also refers to 18-wheeler

Running Barefoot – CB *without* an illegal linear

Shaky side – West coast

CB Glossary, continued...

Step on – A higher power transmission blocking another, as in "He stepped on me!"

Threes – Regards

Three-three, or thirty-three – emergency

Tijuana taxi – Fully painted and lit police car

Twenty – 10-20, location: as in "what's your 20?"

Uncle Charlie – FCC, including monitoring trucks

USB – Upper sideband

Vizwar – VSWR, voltage standing wave ratio, interface efficiency between two radio-associated devices such as transmitter and antenna

Wall-to-wall – Maximum transmitting power

Wrapper – Paint color, as in "Smokey's in a tan wrapper."

What will you get from monitoring CB? Beside a potential headache, and human insights that would make Freud blush, you'll get the location and character of...

- the worst and the most cost-effective restaurants
- the female servers
- roadside bars
- "social clubs"
- neighborhoods where ladies of the evening take the air
- habits, tactics, strategies of local law enforcement
- nearby radar speed traps, and their schedule
- dangerous intersections and areas
- low cost fuel stops
- cost-effective rooms and motels

...and much more. Many good buddies have adopted portions of the law enforcement "Ten-Code," and even added elements of their own invention.

Listening to CB is frustrating, also, because the band is full of overlapping near and distant radio transmissions and the one you're listening to will often be "stepped-on" by a stronger one. To "punch through" requires an every stronger "foot warmer," and a good percentage of CB communication deals with illegal equipment.

Power and Frequency "Flexibility"

A conventional/legal Citizen's Band radio can transmit no more than four watts of Peak Effective Power (PEP), which becomes twelve watts in single sideband (SSB) mode. That's power enough to communicate "line of sight," or even slightly beyond because of bending effects at CB frequencies. It is not sufficient power for the CB aficionado, who installs a "foot warmer" (linear amplifier) with hundreds or even a thousand watts of power. Yes, it's illegal. Worse, it saturates a given frequency, or channel, for a great distance.

When 40 people within a few miles all turn on linear amplifiers on 40 channels, there is no Citizen's Band. If everyone with a linear amplifier were to junk it tomorrow (thus clearing up contamination of local communication by distant linears), overall communication would probably improve even for those who abandoned their illegal equipment.

There are yet more opportunities for a CB operator to go past the law. A crystal-controlled CB uses either a set of 40 crystals plus a switch, or a single crystal plus a synthesizer to generate the desired 40 channels. By replacing the crystal to one at a new reference frequency, all the channels are shifted to non-CB frequencies. Such users shift to frequencies just above or just below the CB band, where the absence of radio signals permits long-range and clear communication.

For this scheme to be of any use, there must be multiple operators working together at the new frequencies, though it's not uncommon for a CB base (at home) and mobile (in the family car) to be tuned that way. Illegal, and there's something conspiratorial about this when a group of users get together. The most popular "off the map" frequencies are above Channel 40, in an unassigned (and therefore quiet) area between 27.410 and 28 MHz. Just below Channel 1 (29.965 MHz) is another lightly used area often exploited by illegally modified CB rigs.

Public Safety

CB is obviously fun to use, and almost as much fun to monitor. It also has a practical value. Operation on Channel 9 (the mutually agreed-upon emergency channel) contributes to the safety of our highway system, and that rule is generally respected despite the fact that most CB'ers ignore most regulations that govern their communication. Even CB tends to straighten out in a major emergency. Truckers present during the recent San Francisco area earthquake observed that after a few minutes' confusion, most CB'ers shifted to a helpful and truly professional style, and CB became a resource to the emergency services as well as the CB'ers.

Because of the number of CB radios on the road, a scanner prepared to monitor CB communication during an emergency or disaster can quickly get a broad picture of what's going on.

SSB

Some CB'ers are using SSB modulation, though equipment able to handle it is expensive. SSB improves performance and many advanced CB'ers use it, but only a few scanners have a sideband capability and therefore that part of CB is not available. On the other hand, the CB'ers on SSB tend to be more professional and less amusing.

What *is* available to a conventional scanner is entertaining and worth the time. Like a circus or a horror movie, monitoring CB can be enjoyable – once in a while.

INDUSTRIAL RADIO 12

Usage

Industries use conventional radio to meet mobile non-cellular general communication requirements, and for security, process control, transportation management, emergency services, facilities maintenance, and many other functions that vary from one industry to another. Generally, such functions require applications to the FCC for assignment of the required number of channels. These applications, and the frequency allocations that follow, become public knowledge and can be viewed by any citizen with the interest and will to work through the FCC's administrative processes.   •

The general bands within which industrial radio operates are identified in Chapter 8, though specific assignments vary from area to area. Some of the most interesting industrial users are discussed below.

Utility companies (water, power, gas, cable TV, sewer management, telephone, etc.) all communicate by radio, and locally-published frequency books include their assigned frequencies. It's boring listening until there's a problem, at which time these groups mobilize their personnel and electronic resources and jam the airwaves. Sometimes it's handy to have a few of these frequencies programmed, or at least handy for rapid insertion. For instance, when a utility like the telephone system or cable TV dies, it's usually not possible to get information by phone, either because the phone is dead (though it might be the first thing we reach for to *report* the malfunction) or there's a constant busy signal at the other end.

Set up properly, a scanner can provide early notice of repair efforts, and can even warn of possible associated problems yet to occur as a result of natural disasters such as earthquake, flood, or storm.

Telephone company maintenance personnel use many of the same frequencies nationwide. Some are (all MHz):

43.16	451.175	451.325	456.175
151.985	451.225	451.400	456.675
158.34	451.275	451.625	

Some general federal allocations have been made to highway maintenance. They are (all MHz):

45.64-45.88	150.995-151.0	158.985-159.2
47.02-47.4	156.1-156.24	

Each of the major dams and power production facilities has its own frequency allocations, easily found in appropriate frequency books.

Mobile non-cellular general communication refers to vehicles equipped with industrial, rather than cellular, radios. Whether generated by a pickup truck delivering parts from a warehouse to a repair facility, or a staff driver awaiting a passenger at the airport, communication over industrial mobile radio is usually boring. Radio transmissions by security agencies, on the other hand, point to where the action is, and you'll read more about the security business later.

Industrial "process control" communication is intended to support or improve production or process management in oil wells and refineries, farms and ranches, quarries and mines, and similar industries where a shout is not nearly enough. Cropdusters, ranch hands, toxic substance control people, and others all use the radio in these bands. It's always interesting to just listen, as new names and allocations rise and fall with the steady pendulum between venture financing and the bankruptcy judge.

Equipment

Industrial radios are made by Motorola, GE, Bendix/King, and similar manufacturers. They're either installed in a vehicle or handheld, and racks at some convenient point permit the batteries to be recharged. Some operate on only one frequency or one pair of frequencies only, and therefore use one or two crystals, and to change from one channel to another often requires partial disassembly. Such industrial communication networks will often use the same frequency assignment for many years, without variation, which often makes for saturation as the user grows. Other radios, far more flexible, use a single crystal plus a frequency synthesizer. They cover the desired operating band, and can be tuned to any of the frequencies within it. Obviously, that is a much more useful – though more expensive – solution.

These radios are subject to very difficult environments and operating conditions, and are dropped often, get wet in the rain, and are often abused. It is a tribute to their manufacturers that they are so durable. Some are even designed to be absolutely sealed and spark-free ("intrinsic safety") so they can be used in environments that are subject to explosive or combustible gases or liquids. Of course, there is a formula that takes into account required durability and required cost, and that usually prohibits scrambling. The absence of even the most primitive scrambling in the industrial communications business makes scanning and monitoring very easy.

As an MBA student learned while collecting information for a thesis, a scanner might tune to conversations you'd rather not have heard. This "legend" goes as follows: she was working on a study dealing with the oil industry, and parked near a gasoline distribution point. A pipeline delivers fuels to that market from a refinery about 100 miles away, with different petroleum products separated by traveling "plugs" that prevent contamination from one "shipment" to another. She learned that trucks which service the major brand gasoline stations get their loads from the same pipe.

In other words, she overheard truckers from various oil companies all hooking up to the same tank (identified by number). The implication is that the difference between brands might be nothing more than neon and advertising budget, though we would like to think that at least a different packet of additives was added to each brand of gasoline.

Frequency Allocations

Some specific federal allocations to industries are listed below (all in MHz, and all are within 0.5 MHz, so scanning is required):

Petroleum	30.66-30.8, 33.1-33.4, 48.56-49.6, 153.05-153.68, 158.2-158.46, 456.175-456.750
Power/Utilities	47.7-48.54, 74.6-75.2, 153.4-153.68, 158.1-158.28, 451.00-451.7, 456.0-456.675
Telephone	451.175-451.70, 456.175-456.675
Manufacturers	72.0-72.5, 153.05-153.95, 158.28-158.4, 462.175-462.525, 467.2-467.525
Motion Picture Production	152.87-153.05

Many of the bands designated "land mobile" can contain frequencies assigned to various industrial users. In addition to the general allocations listed above, one can search for industrial users near the following frequencies (all in MHz).

35	57	151-154	461-465
43	72	452	856-865

Local frequency books generally include frequency allocations to industry, but not all are always listed. One of the easiest ways to identify the frequencies used by a particular organization is to simply ask first the operator and, if you don't get a satisfactory answer, the FCC. Some operators believe the frequency is a sort of secret, and others recognize that the FCC is generally required to provide the answer to that question. Be polite – petty bureaucrats need to feel important.

Forestry

It may seem that an overly large segment of the spectrum is allocated to "Forest Products," but there's good reason. Forestry is one of this nation's largest industries, and except for fishing it is spread over the largest area. The terrain is often difficult, transportation is extremely limited, and population is sparse. Communication is therefore critical, and because our forests are such a precious resource, fire emergencies multiply the importance of radio. The FCC has therefore dedicated wide bands to forestry, with justification. Among them (all in MHz) are:

31.0-31.6	157.56-157.74
43.0-43.52	158.2-158.4
48.56-49.6	451.175-451.750
72.44-72.6	452.1-452.325
151.145-151.49	456.175-456.675
152.48	457.1-457.45
153.05-153.68	

Most ranger stations and observation towers are on high ground, so you can probably monitor their transmissions if the mountains are visible in the distance. A detailed map helps, and many are published with notations indicating the locations of such towers. If a military map is available (or a color VFR map from a local flying club or airport), your estimates can become quite precise. With patience and perhaps a few phone calls you can determine the call signs of some of the towers, and then it gets interesting.

When a ranger spots a plume of smoke, other towers are called, azimuths (compass headings to the fire) are taken by at least two towers, and the cooperative effort supports triangulation to determine the exact location of the fire. You can then overhear vehicles entering the area, and occasionally even an airborne assault. Often a small aircraft will circle overhead to direct the paths and drop points of the slurry bombers, and it's quite exciting to listen to the deadly war against the flames. Forestry services have a lot to offer the scannist, and though fires are bad news, that's too frequently what's on the air. One dry summer listening to the firefighters and you'll become especially careful of your matches.

Wheeled Ground Transportation

Trucks, vans, taxicabs, limousines, and buses all use various industrial and land mobile communication bands for coordination and reporting.

Truck communication is divided into two geographic categories and three radio categories. Local trucks typically transmit voice near the railroads, at about 160 MHz, and use local repeaters to ensure communication throughout their operating area. Cross-country truckers use CB for fun and Bear avoidance, and a growing number of trucks are equipped with satellite transceivers that permit position location (through Loran or the Global Positioning System, or GPS) plus reception of printed messages such as "Pick up another two tons of gold bullion at Gondorf's, 35 miles further on I-15." While such text messages can often be intercepted with the right equipment (though they're digitized), they're of little interest unless you're a hijacker or your name is Gondorf. The government has allocated specific frequency bands to the "motor carrier" industry, as follows:

3.7-44.62   159.48-160.215   457.325-457.8
30.6-30.9   452.325-452.95

Taxicabs and limousines are sometimes fascinating, but there are usually hours of chaff between kernels of wheat. In some urban communities there is a certain connection between some limousine services and prostitution, which spices things up, but if that's of real interest it's easier to go to the phone book.

Most federally-standardized taxi frequencies are between 152.27-152.48, 452.05-452.45, and 457.1-457.45 MHz. One can learn a lot about a city by monitoring the taxi drivers for a while. It takes patience.

Buses are another matter. Between breakdowns, accidents, angry passengers, onboard altercations, and observation of unusual events, the bus drivers of our urban areas are often entertaining and informative.

"Trunking" was developed and approved by the FCC primarily to improve the efficiency of spectrum usage, and particularly to support "dispatch" operations, which includes the control of industrial transport and delivery vehicles. So don't be surprised if your scanning objective in that category is a trunking subscriber. Some of our following topics involve trunking, so get your thumb ready...

e. e. cummings
One Times One (1944)

"Listen: there's a hell of a good universe next door; let's go listen."

A great observation, and here's another by old "e.e." that has nothing to do with scanners, but is too good to miss:

"A politician is an arse upon which everyone has sat except a man."

NATIONAL TRANSPORTATION AGENCIES 13

Railroads and Subways

Railroad fans are a dedicated bunch. They must be, since there is little to say about trains, and such transmissions are line of sight so they must be nearby or repeated to be heard. Imagine how exciting it can be to be the first in your neighborhood to learn that a particular train will be passing through it within an hour. That's enough time to tie up Pauline and carry her down to the track! Each to his own taste. Those who spend uncounted hours listening to railroad communications defend their interests and their hobby, and are safer there than throwing darts at the local pub.

Such listeners have something in common with the practitioners of the world's most boring hobby, in England, sometimes called "Watching," among other things. There are literally hundreds of different types and classes of rail equipment in use in the United Kingdom, and though schedules (pronounced "shedules," of course) are published they list arrival times but do not identify the specific equipment destined to meet the shedule.

To Watch, one goes to a local railroad station carrying a book that lists every type and class of train and engine on rails in England, and when one identifies a particular piece of equipment for the first time, the appropriate point in the book is check-marked. One might spend days at the station before seeing anything not already checked, particularly when off the more heavily traveled paths.

When the book is filled, what next? Well, the hobbyist buys a new book and starts again, of course.

U.S. railroads generally operate from 160.215 to 161.565, and in some areas there are repeaters that extend range over the horizon. Railroad buffs are upset to have discovered that the FCC may move their hobby to another frequency range, as the 160 MHz band will be allocated to other applications.

Humor aside, listening to railroads and subways is popular – and even fascinating at times – and there's a significant degree of organization in this segment of the scanning hobby. For instance, it is supported by computer bulletin boards that provide news on frequencies used by various railroad transmitters. The *RCMA's Scanner Journal* publishes a regular section on railroad scanning, which is complete, frequently updated, and generally well done. A typical news item might be, *"Jack Smith of Dallas reports that Road Channel 3 is 160.385, and that's what they're using on the Long Branch to Whitney Junction."*

A periodical called *The Capitol Hill Monitor* lists frequencies for the Maryland and DC railroad systems, and even covers the Washington DC Metro. The railroad scanning hobby includes a lot of *underground* activity, too. New York uses radio to monitor subway activity and security, as do most other subway systems in the Northeast, and most of those frequencies have been published.

Some railroads, particularly within metropolitan areas, are transitioning to cellular communications by negotiating rates with local cellular carriers and buying commercial equipment. It's an economical shift, and has the singular advantage of shifting maintenance of the system to another party. When the last cellular-capable scanner breaks down irreparably (scheduled for August of 2120), the transition to cellular will provide communication security to railroads for the first time.

If you have patience and a certain "attitude," you might decide to scan the railroad frequencies, but never with the hope of hearing "Quick, Jack. Turn right!"

Maritime

Ship-to-ship and ship-to-shore communication use many different bands, including (all in MHz):

156.2475 - 157.45	454.0 - 455.0
161.575 - 161.625	459.0 - 460.0
161.775 - 162.0125	(private intership, at sea)

These are line of sight radio links, and are useless when the ship is more than a few dozen miles from land. Because they are line of sight, however, they rarely interfere with similar frequency allocations to landmobile operations.

For long-range communication from ship to shore, the industry uses HF – the band from 2 to 30 MHz – and the customary modulation is single sideband (SSB). HF, even SSB, is characterized by "fading channels" that generally require constant attention and retuning by either a computer or a trained radio operator. While almost all quality short wave radios can tune such signals (with manual assistance), ordinary scanners cannot.

Aviation

Both the commercial and general aviation industries are fine generators of interesting radio communication. Though they generally use the same frequencies, they differ in purpose, equipment, priority, training, and language. There are excellent books published on aviation communication protocols, frequencies, stations, and language. The reader is encouraged to use this chapter as an introductory guide, to determine whether the topic is sufficiently fascinating to justify buying one of the books that focus on aviation.

The majority of radio communication from aircraft is with controllers on the ground, though a small amount falls into other categories such as mechanical problems, alerts to maintenance staff, advice by the airline's ground staff of gate assignments and schedule changes, radiotelephone conversations by passengers, administrative interchanges concerning flight assignments of the crew, and warnings by the airline's security staff of bombs, threats, etc.

The Federal Aviation Administration entities with which aircraft communicate include ground control, approach control, departure control, and air traffic control. Though it's beyond the scope of this book to conduct a comprehensive tutorial on aviation language, it's not fair to claim that listening to local aviation is an interesting application of a scanner without providing some insight as to who's doing what, and just what the jargon means.

Ground Control is responsible for aircraft that are taxiing from their gate to the runway, or back. They monitor aircraft on the ground, coordinate with approach and departure controllers to ensure that aircraft crossing runways can do so safely, and interact with airline or (in the case of general aviation) air service facilities to ensure support and coordination. Ground control conversations sound about like this:

ATC clears N12345 to the Midland Airport via Victor 33 to Argo, Victor 44 to Midland. Climb to and maintain flight level 220.

For this flight, Air Traffic Control has established a route from take-off to destination via airway #33 to an intersection of two airways called "Argo," and then via airway #44 to destination. The clearance acknowledges appropriate separation, both vertical and horizontal, from other aircraft under the control of the same agency. The aircraft is instructed to climb to 22,000' above sea level.

N12345 is cleared to taxi to Runway 36 via taxiway 4. Hold and call before crossing Runway 27. Barometer 29.92, winds north at five, ceiling 300, runway visibility one mile in fog.

The aircraft was directed to taxi to the south end of the north-south runway, via the #4 taxiway from the parking area. It is not to cross the east-west runway without express permission. The atmospheric pressure is neither higher nor lower than average for sea level. Winds are blowing from the north at five knots. There is a visual obstruction 300' above the runway. Horizontal visibility is limited to one mile because of fog.

and then...

N12345 is cleared to take position and hold. Switch to Departure Control on 120.5. Monitor Guard. Good day.

The aircraft may taxi onto the south end of the north-south runway, facing north, and await permission to add power and begin moving. The VHF radio is to be switched to 120.5 MHz and communication established with Departure Control. The 121.5 MHz emergency channel is to be monitored.

Departure Control manages the aircraft from the time it takes position on the runway until it is safely out of the airport control zone, when control is passed to Air Traffic Control. Departure Control communications sound like this:

Delta 123 is cleared to take off. After take-off turn right to zero-four-five. Climb to and maintain 3,000. Report Point X-Ray outbound.

The aircraft is cleared to apply power and begin rolling. It is to initiate a right turn to the northeast. It is to climb to 3,000' and level off pending further instructions. When passing a radio beacon or a geographic point called "Point X-ray" it is to inform Departure Control accordingly.

Approach Control takes responsibility for the aircraft from the time it enters or approaches the airport control zone until it is safely on the ground. One subset of approach control provides Ground Controlled Approaches (GCA), a radar-assisted landing usually used in bad weather. It sounds like this:

Marine 12345 is cleared to hold at Point X-Ray at 5,000. Standard turns, three minute legs.

This Marine Corps aircraft has been instructed to fly to the radio-defined Point X-Ray. It is then to maintain 5,000' and establish a racetrack pattern over the ground, with right "U-turns" and three minutes between them. Airspeed is understood to be optimum for endurance.

Marine 12345. You're next in the pattern. Cleared to Point Zulu at 5,000. Report X-Ray. Contact GCA on 123.5 if able, else 234.5. State?

12345 may leave the holding pattern the next time it reaches Point X-Ray, and is to report that point. It is cleared to go from X-Ray to Zulu. It is not cleared to descend. It is to switch to radar control on the VHF frequency, but since few military aircraft are equipped with VHF a UHF alternative was provided by a knowledgeable controller. "State" is a question, asking fuel remaining.

The tower itself (that tall building with the tinted windows around the top) can get into the act when the weather is suitable.

N23456 is cleared to Elliot Airfield via flight planned route. Maintain VFR. Monitor Guard.

The tower has given this aircraft a Visual Flight Rules (VFR) clearance to fly from takeoff point to destination. VFR implies sufficient visibility enroute to permit the crew to take responsibility for separation from other aircraft, and the flight must be conducted at an altitude that corresponds to VFR. N23456 has been reminded to monitor the emergency channel.

Commercial aviation includes common carriers, such as the airlines with which most are familiar and the cargo carriers that may be new names to some readers. They carry cargo and passengers in accordance with a regular published schedule. In dialog with ground control agencies, pilots identify themselves with a call sign that includes the carrier's name plus the flight number. Crews of widebody aircraft such as the Boeing 747 and the Douglas DC-10 add the word "heavy" to alert controllers to their weight, turning radius, and inertia, and to remind them that extra ambulances will be required if they make a mistake.

"General aviation" refers to that veritable cloud of Piper, Cessna, Rooney, and similar aircraft that buzz around smaller airfields throughout the country.

As indicated above, military aviation enjoys full privileges at most airports, and interacts with control agencies in the same standardized manner as general aviation and commercial aviation.

Commercial and general aviation aircraft, and the agencies that support them, all operate on VHF from 108 - 136 MHz and monitor Guard on 121.5 MHz. Military aircraft are in the UHF range from 225 - 400 MHz and monitor Guard on 243.0 MHz, and are also found in the VHF range from 118-140 MHz.

To properly enjoy scanning aviation activities, the hobbyist should purchase charts and approach plates from a local flying club. Charts will show the local airways, airports, control zones, and restricted areas. Each approach plate depicts landing patterns for a specific airport, including details of radio beacons and a drawing of the airport layout.

If you're near the airport, airline ground communication can be a lot of fun. No, it's not the dialog between ground control and the pilot over which taxiway to use, but the usually interesting and often hilarious discussion between the aircrew and the airline staff when the aircraft is on the ground or at least near the airport. It's almost all on UHF, near 460 MHz, and scanning in that range, during operations, will quickly identify just who is using what. It helps to have a pocket air schedule handy so the flights can be identified by number as they land and taxi.

Airline Private Frequencies

Airlines have been allocated frequencies for administrative purposes, also, to support maintenance, schedule, and VIP problems. Here are some known users and frequencies, all in MHz:

Airline	Frequency
Air California	460.7
Alaska Airlines	460.825
American Airlines	460.7, 460.775, 462.8
America West	464.6
Continental Airlines	460.7, 464.875
Delta Airlines	460.75, 460.825, 462.725, 464.575
TWA	460.675
USAir	461.6125, 461.9625, 464.475, 464.875
United Airlines	460.725, 460.875

Scanning the aviation world is interesting, and if it's supported by charts and plates it's educational as well. Frequencies in use at airports are in the public domain, and are generally available from locally published frequency books or from the communications manager of the airport.

MORE SCANNING 14

Sports Events

Virtually every track, stadium, and team has its own frequencies. Whether your sport is football or ping-pong, a scanner can add to the pleasure of a good competition.

You'll have to dig up the frequencies for the facilities in your area, but for the purpose of this book Indianapolis sets a fine example, as shown in the *Scanner Journal* (see Bibliography) which regularly publishes frequencies for sports events. Taking a scanner to (or near, for that matter) the Indianapolis Motor Speedway can only enhance the pleasure of the annual 500. Race officials, teams, security, vendors, transporters, and other participants all have their allocated frequencies, and virtually all can be scanned. Here are some of the frequencies of that particular facility, all in MHz.

Pace car	452.275, 935.1375
Ambulance	467.75
U. S. Auto Club	151.625, 151.655, 466.6875

The National Hot Rod Association uses 154.54 for staging, 151.625 for safety crews, and 154.570 for administration. The National Off-Road Vehicle Association uses 151.625 also, plus 151.925. NASCAR uses 464.500, 464.775, 464.900, 469.500, among others. Each driver in the Winston Cup had a frequency allocation, near 467 or 852 MHz. Such allocations operate at raceways across the country.

Football teams all use radio now, as do baseball and soccer teams, and it's not only legal to monitor them, it's fun. Again, the best sources by far for such frequencies is your local frequency list and such guides as the *National Scanning Report,* the *Scanner Journal,* and *Monitoring Times.*

Universities and Colleges

A typical university is like a small city, with residences, industrial sites, medical facilities, transportation, security problems, and more. Therefore, it communicates like a city. The preponderance of such communication is over the phone, but radio traffic is brisk and even sometimes interesting.

Interest peaks during sporting events or when a dignitary visits, because security is activated and logistics get complicated. A detailed map of the university is a must if scanning is to be meaningful and at all entertaining.

No frequencies have been assigned to universities nationwide, so the hobbyist's best resource is either a local frequency book or the communication manager of the university (who is often within the security department, and therefore a bit paranoid).

Hospitals

Hospitals are also somewhat like a town, though they have many emergencies in comparison. It's useful to acquire a local map and helicopter/ambulance/paramedic frequencies if what's heard is to be meaningful.

The frequencies can be found in a local frequency book or by contacting a hospital's cooperative communication manager.

Security Services

Security services are fun to monitor, particularly if there's a large organization in the area. Their communication is typically on one or two frequencies in any given city, and monitoring those frequencies can generate early information on events ranging from bank robberies to break ins. Of course, the preponderance of the communication concerns shift changes and coffee breaks.

In at least one major urban area, Pinkerton's uses 461.825, 464.2, and 151.655 MHz. A nationwide alarm company, ADT, operates reaction trucks and uses 460.950, 461.175, and 464.725 MHz in some communities. The best bet is to refer to a competent frequency book for the area of interest, as most include security services.

That's also the easiest source for the frequencies used by the security forces of shopping centers and malls. Virtually all such forces use radio to coordinate their efforts, and to communicate back to a central dispatcher/controller who can then call upon law enforcement or fire department resources as required.

If the security of your money is interesting to you, consider the boring hobby of listening to armored trucks as they move from bank to bank, and store to store. In some urban markets, Armored Transport International uses 452.775 MHz, Brinks uses 44.2 MHz, Loomis Armored uses a trunked system in the 800 MHz range, and Wells Fargo uses 461.050 MHz.

Note. Virtually all states have passed legislation that makes it a criminal act to intercept certain security and law enforcement transmissions and divulge them to perpetrators of crimes. Think about it. Some scanner hobbyists have been warned off for foolishness, but deliberate support of criminal activity can be a very serious offense. Let the bums buy their *own* scanners.

News Media

News gathering is a complex task that involves vehicles (often with satellite link radios and dish antennas), people on foot, and communication links with law enforcement agencies and other emergency services.

The ground-to-ground radio links used by the news media are

published in the various frequency books. While the majority of the most useful "raw" (unedited) information travels by satellite link to the studio, some of the most fascinating "early warnings" of breaking news are available to the scannist.

Of course, one can correctly assume that most reporters use scanners to alert them to breaking news. The more avid hobbyist should listen to the same transmissions the reporters tune, and get the news even sooner. During the 1994 O.J. Simpson saga, scanners came into national prominence (again) when all the media and thousands of private citizens used them for second-by-second news. Specific federal allocations to press relay services are near 452.95 and 457.95 MHz. Microwave relay (using a dish antenna) is generally not accessible by scanners.

Pagers

The pager business is not ordinary... First, it's a consumer electronics product the production of which is dominated by Motorola's Florida operation. Second, it's a service industry comprised of thousands of small businesses. Each acquires a license, buys a pager transmitter and some pagers, leases lines from the phone company, places advertisements, and that's that. Many of these businesses are truly entrepreneurial. The privacy of the service they provide is jeopardized by scannists who simply *must* climb Mt. Everest, just because it's there. Here's how pager privacy is violated...

Pagers receive a digital data stream that includes the specific digital address of that pager, plus the number to be called and, in the case of alpha-numeric services, text. Decoding the text is a simple task of computer programming, and some companies actually sell the required software.

Voice paging makes more sense to the insatiably curious, and this is a growing business. Such transmissions are broadcast within 0.5 MHz of 35.5, 43.5, 152.5, and 454.5 MHz, and some more recently established operations are now in the 900

MHz band. Voice paging usually dictates that the caller make a point quickly and emphatically, so calls range from business appointments to love notes.

Evangelists

One of the most fascinating industries in the United States is the evangelical movement. Evangelists are assigned their own radio frequencies, though it may disappoint them when their communication equipment works only with terrestrials. Most such communication involves setting up and managing the evangelical function, providing security and transportation, and handling a host of logistics issues. On the other hand...

The December 1992 issue of *Popular Communications* featured a fine article by Chuck Robertson, which included a sizable list of televangelical movements and the frequencies assigned them by the FCC. For a complete list, find *Popular Communications* in the Bibliography and contact the publisher. A few of the frequencies provided by Mr. Robertson are reprinted on the next page. Sometimes the name of the televangelist is used, though the actual licensee may be an organization.

Tuning into an evangelist is not like a revelation by a divine being, but can produce almost as profound a change in the listener's attitude, especially toward organized religion. Scanning a cult's security communications is an equally iconoclastic experience. One "deprogrammer" used a scanner and spent considerable time outside a cult's camp. He claimed that eventually he gained a reasonable picture of what went on inside, and that the information was useful.

Eavesdropping upon the conversations of the "insiders" of *any* religion can be truly enlightening, but not always in the manner they seek, especially when they don't know there's an audience. Somehow, members of this industry seem to have faith that no one is listening.

Televangelist Frequencies

Name	*Frequencies*
Jimmy Swaggart Baton Rouge, LA	463.825, 464.5, 467.85, 467.9
Paul Crouch Tustin, CA	464.8625, 464.9125, 469.8625, 469.9125
Jerry Falwell Lynchburg, VA	154.515, 450.65, 464.775, 947.0, 947.875
Marilyn Hickey National Frequencies	173.325, 173.375
Billy Graham Minneapolis, MN	461.625, 466.65
Oral Roberts Tulsa, OK	461.65, 461.15, 461.4875
Robert Shuller Garden Grove, CA	154.57, 154.6, 154.57
Sun Myung Moon Irvington, NY	464.375
Church of Scientology Clearwater, FL	464.775, 468.3375

Those who wish to scan this select group should be prepared for a bout of serious iconoclasm. Listening to the evangelists is *not* for the utterly faithful, and if you were about to reach for your checkbook to write out a "donation," perhaps you'd best grab your scanner first, and tighten your seat belt!

The Amazing (James) Randi, a magician and famed debunker, taped such radio transmissions for use in his book *The Faith Healers*. That's *not* the book to have on your coffee table the next time a passionately-converted neighbor drops by.

TELEPHONES 15

Air-Ground Telephone Communication

Jack Goeken formed MCI (Microwave Communications, Inc.) to cost-effectively transmit information between two midwest cities, and the firm eventually grew into a major competitor of AT&T. Later in his career he developed Airfone, which appears in most of the country's air lines today. Whether on a wall or in the seatback, Airfone is activated by a credit card and permits limited air to ground telephone conversations. Airfone was sold to GTE, and after a suitable period Mr. Goeken created a competing company called InFlight Phone Corporation.

Today there are several companies offering to the airline passenger not only telephone communication, but entertainment, interactive games, stock market quotes, reservation services, shops-in-the-air, and much more.

The system depends on a multitude (perhaps up to 70) of complex earth stations, each equipped with radios, a computer, and links to the ground telephone system. The original Airfone would seek the strongest station and establish communication through it, but when the aircraft flew out of range the signal simply went dead.

More modern systems from Hughes Avicom and InFlight, among others, operate much like cellular phones. When the aircraft moves from one area to another, the aircraft radio switches from one ground station to the next. An obviously better solution, though it requires communication between earth stations and much more computing power.

In addition, some systems utilize satellite communication to augment the ground-based system. See *Figure 34*.

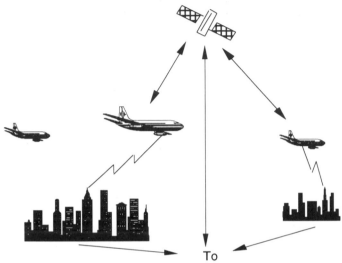

To
Plain Old Telephone Service (POTS),
for connection anywhere phone service goes. The link also
goes to a master computing and data source that transmits
entertainment, news, sports, etc. for passenger enjoyment.

Figure 34

While the original Airfone was awarded an FCC Experimental license to operate over narrow bands in UHF, newer contenders have filed applications for other bands. Generally, air-to-ground telephones use 849-851 MHz for the earth station transmitters, and 894-896 MHz for the aircraft transmitters. Channel spacing is generally 6 kHz (for voice). Modulation can be FM, SSB, or the newer systems can use digital techniques beyond the ability of a conventional scanner.

Current technology permits calls from the aircraft to the ground telephone system, but not in the other direction. That may change when more computing power is added and passengers are permitted to "register" when they get aboard the aircraft; that will permit calls to be placed to the aircraft, and even to the specific seat.

Some firms are using their airline installations to justify the investment to implement the expensive earth stations, and once done are turning their attention to private aircraft. Within a short period, executive aircraft and even single-engine planes will have telephones aboard that operate very much like cellular phones. It's getting harder and harder to hide from the office... or one's mistress.

Though air telephone systems are now built and operated by several companies, and appear on most airlines, the installers are facing an uphill battle because of poor performance of older systems, and high cost. Passengers therefore use them rarely, and while additional features and voice quality will gradually convince the market, for some time to come the listener will find scanning air-to-ground telephone communication generally unrewarding.

Is it legal to listen? Well, the author has found no specific federal legislation that prohibits listening to air-ground communications, and the first-generation systems (Airfone) are neither digital nor encrypted, so if you feel compelled to hear how people make hotel reservations and rent cars, there might not be much legal risk.

Cordless Telephones

About 41% of American households use a cordless phone, so assume that in a modern apartment building of 100 units, perhaps 41 such phones will be installed. If the average household conducts five 10-minute conversations a day on that equipment, that's 205 conversations a day for a total of more than 34 hours of cordless dialog. Such calls wane during the early morning hours, so the odds are pretty good that at any given moment from early morning to midnight, there are at least one or two conversations being conducted in that building. If one percent of adult American men own a scanner that can tune to the range where cordless phones work, the odds are also good that those calls are monitored by a stranger.

When you make a call using a cordless phone, both sides of the conversation are easily available to anyone who is curious, properly equipped, and within range. Nearly all scanners can tune cordless phone frequencies, and even with the standard rubber ducky antenna, nearby calls will be loud and clear. With a tuned whip, reception is even better. Under good conditions, a high quality scanner wired to a rotating TV antenna can pick up cordless telephone conversations more than a mile away.

Cordless telephones should *never* be used for sensitive dialogs. *Never!* They're just too easy to monitor, either with a scanner or with another cordless phone, and there are just too many people with nothing else to do but listen in. There is absolutely no way to know whether such a listener is out there, so always assume there is one. Want proof? Periodically, call an "inside" friend, excitedly claim that "someone's breaking into a car outside," and go to the window...

Prior to 1984, cordless telephone *handsets* transmitted in the 49 MHz range, and the *base* used various frequencies below 1.8 MHz. Phone systems built after that period use 46 MHz for the base and 49 MHz for the portable handsets. This equipment uses common FM, and is easily demodulated. In fact, any cordless handset can usually listen to any nearby cordless communication in the same band.

The exception to the familiar frequency planning and modulation of cordless is called CT2 by the FCC, which means Cordless Telephone 2, planned as the next generation. Like many "next generations" in electronics, CT2 is digital. The FCC opened the band from 902-928 MHz for personal communications, and many entrepreneurs moving to that new segment of the spectrum are using digital modulation schemes.

In 1993, the market witnessed the emergence of several new cordless phones that operate in that range, using intrinsically encrypted digital modulation. The first to enter the market was built by Video Communication of Hong Kong, and marketed by VComm, in Beaverton, Oregon. In this country, the phones are sold under the brand name *Tropez*, at prices up to $300. They use bit-stream encryption, which reduces any chance of casual eavesdropping, and because little interference exists in the 902-928 MHz band, and the digital protocols include error-detection and error-correction, the range of the Tropez is up to 1,000 yards, or more than three times the maximum range of the conventional cordless phone. The Tropez phone is not easily scannable, but an even tougher challenge to the listener is emerging: true spread spectrum (SpSp).

The first SpSp phones to reach the market are by Panasonic, Cincinnati Microwave, Inc. (remember the Passport and Escort radar detectors), Uniden, BellSouth, and AT&T, and by ordinary cordless standards they're all expensive but work so well many of us won't look at the price. Panasonic's entry is carried by consumer electronics stores, whose salespeople claim a mile range. CMI's Escort 9000 will work anywhere in the neighborhood because of its nearly half-mile range. AT&T's phones have even longer range, though they're more costly than the others. These technically aggressive products, and others to come, solve three problems. They're secure, provide crystal-clear audio, and have long range. They solve one other problem. The national market is about 12 million cordless phones per year, and digital designs are quite a bit more costly than their analog counterparts, so SpSp promises to solve revenue problems for their marketers.

SpSp was once a highly classified government technique, alleged to have been originally developed (believe it or not) by WWII pinup Hedy Lamarr and her associates, to permit very secure communication within the military. SpSp simultaneously distributes the "intelligence" (voice and data, as examples) over a wide band of frequencies in the form of a digital code.

In itself, that is interesting and believable.

But even some engineers find it difficult to accept that an SpSp signal can actually disappear within the environmental noise and still be retrievable by a receiver equipped with the correct code. As *Figure 35* shows, ordinary radio signals appear as discrete spikes of energy at specific points in the spectrum. SpSp signals, on the other hand, are spread and buried in the noise, so finding them takes special talents and equipment.

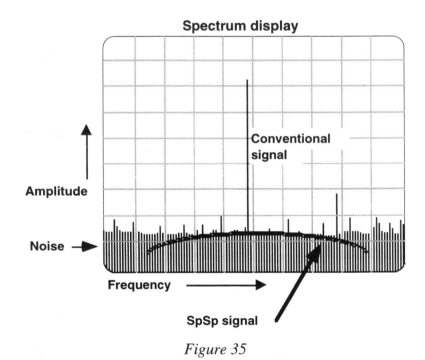

Figure 35

Detecting an SpSp signal is like picking out the voice of your daughter from the roar of a stadium crowd. Most mothers can do it, but it's tough. So far, no one has come up with economical ways to analyze such noise and discover the signal, hence SpSp communication is generally private.

SpSp signals are called "Low Probability of Intercept" (LPI) in the communications trade. If a scanner can't find an SpSp signal, neither can a broadband receiver of the type used by security services to discover "bugs." SpSp is sufficiently private to have emanated from picture frames, memorabilia, and ashtrays in embassies and meeting rooms around the world. It's certainly a resource to intelligence agencies, but less known is its suitability to domestic and industrial surveillance. Though magazines are full of ads for bugs of varying sophistication, only a very few cannot be discovered with a scanner.

And if a transmitter can be discovered, it can also be monitored. And if it can be monitored and recorded, *eventually* the signal can be decrypted. It's a matter of time and resources.

Monitoring Cordless Phones

Note: in some states it may be illegal to monitor conversations on cordless telephones, except under specific conditions and by authorized agencies.

Conventional cordless phones – and wireless baby monitors – use frequency modulation, typically narrowband, and for economic reasons (and because the better FM units are simple and quite durable) they will represent the majority of cordless usage for many years to come. They are extremely vulnerable to monitoring, as are "baby monitors" that use the same frequencies and modulation. Baby monitors are usually left on 24 hours a day, and only the receiver is turned off when it is recharging. If the transmitter and crib are kept in the master bedroom, the unit should be turned off at night.

There are two ways to set up a scanner to listen to cordless telephone conversations and baby monitor transmissions. The first is to simply set a scan range from 46.61 MHz (the lowest frequency used in modern units) to 49.97 MHz (the highest), with steps of 20 kHz (or any submultiple, such as 10 kHz or 5 kHz). A fast-stepping scanner will cover that range in a few seconds, and little will be missed.

The second way is to program the 46.xx MHz channels and the 49.xx MHz channels into the scanner, and then scan through them. Since the channels are so close together, it's about an even choice. Nevertheless, there are ten allocated cordless frequencies, all in MHz, and they can be individually programmed into a scanner if the user prefers.

Cordless Telephone and Baby Monitor Frequencies

CH	Base TX	Handset TX
1	46.61	49.67
2	46.63	49.845
3	46.67	49.86
4	46.71	49.77
5	46.73	49.875
6	46.77	49.83
7	46.83	49.89
8	46.87	49.93
9	46.93	49.99
10	46.97	49.97

Some FM cordless phones are described as having "security features," somehow correlated with multiple channels as indicated above. Such channel selection capabilities have little to do with "security," but reflect the system's ability to find and use that channel with minimum interference or noise. There are scrambled cordless phones on the market, but they're expensive, and the security they provide is more easily attained using spread spectrum.

If you make calls on a cordless phone, remember that there is absolutely no guarantee of privacy even if you're relatively isolated from other structures. If you're in a large apartment complex in an urban area, the odds are very good that someone nearby is listening in.

And if you keep a baby monitor near that crib in your bedroom, turn it off when you're in the room, and especially at night.

On the other side of the antenna, if you feel compelled to listen to cordless telephone communications, remember that though it's sometimes fun, it's usually illegal and can be an unhealthy practice that makes hair grow on your antenna.

The Cellular Pastime

Eavesdropping on cellular telephone conversations is a fast-growing recreation that has captured a significant segment of the scanner market despite the fact that under a 1986 law, it's illegal to do so. An eavesdropper taped a sensitive conversation involving Virginia Governor Douglas Wilder and is alleged to have provided it to a competitor. Another scanner hobbyist taped conversations alleged to be between Lady Di and *her* lover, and between Prince Charles and *his* lover, and that was *not* the same phone call.

A new scanning hobby is growing in some parts of the country. A few eavesdroppers will gather in a well fitted-out van and park in a select neighborhood. The van might be equipped with several scanners, a rotatable directional antenna, and speakers. Each participant will don earphones and scan his own area of interest, and when something interesting is heard on cordless or cellular, it's switched to the speakers for all to enjoy. According to reports, these mobile parties record many of the calls they intercept, and generally have a good time at no cost, because there's a rumor that in Beverly Hills this practice is funded by tapes sold to a select market. Rooftop gatherings in major urban areas, including Manhattan, work similarly.

Scanner hobbyists *can* listen to cellular phone conversations, so they often *do*. It's easy, probably interesting, but illegal.

Monitoring Cellular

Note: it is a federal offense to monitor cellular telephone conversations except under specific conditions, and by authorized agencies.

As this text has shown, most scanners, and even many ham radios, were apparently designed to be easily modified. In fact, the circuit boards of some such units include a loop of wire or a single component that can be reached by removing a few screws. Snipping the component "restores" the ability to tune cellular frequencies. Such simple fixes were found in Radio Shack, Uniden, Bearcat, and other scanner brands.

One argument is that these radios were built for and are sold in countries around the world, of which many have no prohibition against operation in the cellular bands. Therefore, the manufacturers say, it makes economic and engineering sense to build only one design and then make a minor production change for countries in which no bans exist.

A few of the more expensive scanners, specifically by AOR and ICOM, have long been built without any blocking of cellular bands. Their advertising includes the terms: "nothing blocked," or "continuous." Within the next few years, legislation prohibiting or restricting scanners that can cover cellular, and those that can be "readily" modified, will make such units more and more valuable to those who cannot do without cellular phone eavesdropping. In the meantime...

New Legislation

After years of lobbying, the cellular telephone industry was rewarded with legislation to reduce such monitoring, but breaking such laws was always easy. In October of 1992, Congress passed the Telephone Disclosure and Dispute Resolution Act to control abuses of 900-number telephone services. Piggy-backed atop that new law was a new set of restrictions designed to give the cellular industry a persuasive tool to answer consumer questions about security.

Beginning April 26, 1993, the FCC implemented changes to Parts 2 and 15 of its regulations, and will no longer certify scanners that can receive, or be "readily" modified to receive, cellular telephone frequencies. After October 24, 1993, no such scanner could be legally manufactured or imported into this country. Since April 26, 1994, no such equipment has been legally manufactured or imported. Some entrepreneurs have, of course, stockpiled products and now advertise/sell them at a premium.

Also, the law prevents the FCC from certifying mechanisms that can convert digital communication, which is generally immune to scanning, to conventional analog.

As required by the law, the FCC established a procedure that denies certification of certain scanners, and prohibits their manufacture or importation. Specifically, in its January 1993 Notice of Proposed Rulemaking (now adopted), the FCC proposed to deny certification of scanners that (1) are able to tune transmissions in the cellular frequencies, (2) are capable of "readily being altered by the user" to receive transmissions in the cellular band, or (3) can be equipped with decoders that convert digital cellular transmissions to analog voice audio.

To obtain FCC certification of their products, manufacturers and importers must certify compliance, and violations of the new regulations could result in revocation of the equipment authorization, fines up to $75,000, injunctive actions, or even criminal penalties. The cellular industry has trained sales staff to quote sections of this legislation. The objective is to convince the buyer that security is a solved problem. It is not, by a long shot.

This new law places a burden upon the engineers who develop frequency plans for scanners. Most rapid legislative responses to industrial pressure are imperfect, and this is no exception. For instance, while it discusses digital-to-analog conversion, it allows kit-built frequency converters, which can easily be attached to "non-cellular" scanners to enable them to cover cellular bands. On the other hand, it's likely that the FCC and/or Congress will ponderously turn toward reality and pass either a law or an FCC Rule to extend the limitation of the new legislation to cover those devices. That's a pity, since such devices are used for much more than listening to cellular; in fact, there is no other way for an inexpensive scanner to be upgraded to cover the trunking bands in the 800-900 MHz range.

This new legislation had another effect. By making it illegal to produce, or import for sale, cellular-capable scanners *after* certain dates, the implication was that it was perfectly legal to make/import/buy them *before* those dates. Scanner hobbyists who had procrastinated in the past were motivated to get their new equipment while it was available, thus creating a minor gold rush in the industry that began in early 1993. It isn't taking long for various converter "kits" to enter the market, thus allowing the hobbyist to *build* equipment that is otherwise illegal to sell.

Judging by the number of companies that legally sell "test chips" and "test modules" for cable TV descramblers and converters (despite the fact that there's really nothing to test except the user's ethics), and the surprising discrepancies between the evident intent of federal and state lawmakers, it will be tough to get our legislators to coherently produce an enforceable set of laws. Meanwhile, here's what those who choose to (illegally) monitor cellular calls might do...

Assuming the radio was modified or has unrestricted coverage, channel spacing must be either 30 kHz (the standard for North American cellular) or some submultiple of that number. 5 kHz will work, though the radio must scan six frequencies to reach each cellular one.

As an example, suppose the channel spacing is 869.000, then 869.030, 869.060, etc. A radio that tunes in 30 kHz steps will scan only those specific frequencies, skipping everything in between them. A radio that tunes in 10 kHz steps (and is incapable of exact 30 kHz steps) will tune 869.010, 869.020, and then will find a signal on 869.030. It will tune 869.040 and 869.050, and then find a signal on 869.060, and so on. It must inspect three times the minimum number of frequencies to find the cellular signal, so regardless of the unit's scan rate, if it does not support 30 kHz steps (25 kHz in Japan, 200 kHz in Europe) it will seem to operate slowly though in fact it's running as fast as it should.

On the other hand, suppose the scanner can tune only in 25 kHz steps. It's an interesting mathematical exercise to determine how many on-frequency "hits" such a scanner can achieve between 869 and 879 MHz.

The scanner is tuned to the specific band used by the cell radios (as the portables only radiate one half of the conversation), typically 869-879 MHz, the best available step size for that market is selected. Push the SCAN button, adjust the squelch and the volume, and that's it.

DTMF

Dual Tone Multi Frequency can play *Mary Had a Little Lamb*, and if your scanner is equipped with a DTMF decoder, you could find out that there are many ways of constructing the tune, including 7 5 1 5 7 7 7. Though *MHLL* is a noble purpose for such expensive instruments, they might also be used to identify access codes, phone numbers, passwords, and the like, recorded from the air.

Voice mailbox "hackers" have made a game of searching through private information. To discover a password they write a computer program that will generate every DTMF number from 000 through 999 in a few seconds, and put it on tape. Another way is to scan, and then record such tones for later analysis with a scanner. It's a sad piece of work, but this book would not be complete without the actual frequency pairs for the twelve keys on a telephone. All frequencies are in Hz:

KEY	FREQ1	FREQ2
1	697	1209
2	697	1336
3	697	1477
4	770	1209
5	770	1336
6	770	1477
7	852	1209
8	852	1336
9	852	1477
0	941	1336
*	941	1209
#	941	1477

The Ultimate Phone Call Monitor...

Legislation prohibits listening to cellular and cordless calls, and decrypting digitally encoded communications. Omnibus (see Chapter 19) can be construed as prohibiting the hobbyist from monitoring satellite communications that happen to include non-digital telephone calls, and Omnibus is flawed and difficult to apply.

If you absolutely *must* monitor calls from the heavens (to ET?), the technique works as shown in *Figure 36*. A C-band dish if you have one, or a high-gain Yagi antenna (not here... go to a ham store and ask), must be carefully pointed at one of the Telstar satellites or any other that is used for telephone relay. The output of the dish is fed to a conventional TVRO (television receive-only) receiver by Houston Tracker, Chapparal, or other manufacturers.

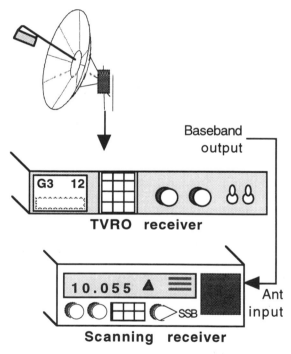

Figure 36

The baseband output of the TVRO receiver is fed to the antenna input of an ordinary scanner or communications receiver capable of operating in the 10MHz range, and having an SSB capability. The scanner or receiver is switched to SSB, and programmed to scan through the 10-11 MHz range. What's out there is a multitude of telephone calls, easily separated with the tuning knob. They come from the ground, where switching offices send bunches to the satellite for relay to a similar switching office across the country.

Most of the calls will be inter-city, but because of the phone system's constant search for routing, some may be between people who are quite close to each other. And because each is holding a wired telephone connected to POTS (Plain Old Telephone Service, a term actually used by Ma Bell), there is an illusion of security even among callers who are aware of scanners.

Check the Bibliography for Harrington & Cooper's book, *The Hidden Signals on Satellite TV.*

PUBLIC SAFETY 16

Emergency Services

Fire departments, ambulance services, and similar emergency services all communicate on bands scannable by hobbyist equipment, and they rarely – if ever – use any form of scrambling. One reason is that an emergency communication is exactly that, and the additional complexity of scramblers adds one more layer of potential failure, which cannot be tolerated.

Local frequency books provide the frequencies used by such services, though most will provide the necessary information if you work your way through the organization to the individual or office responsible for communication. Some of the fire frequencies (in MHz) allocated nationwide are:

35.02	153.77-154.45
33.4-34.0	166.25
45.88	170.15
46.06-46.5	

Monitoring such communication is not only interesting but it can produce potential benefits, particularly during natural disasters and "sociological emergencies," when it's a good idea to know the threat direction and rate of movement so appropriate precautions can be taken.

Law Enforcement

By far, the busiest voice frequencies are those used by law enforcement agencies, which provide the most fascinating information on the air. The reason most people initially buy scanners is to tune law enforcement frequencies; in fact some retailers and manufacturers call these products "police scanners." In any area of reasonable population density, the police are constantly transmitting and the information is nearly hypnotic. The police provide the most interesting communication, the densest (most-frequent, not the dumbest) radio traffic, and in times of social, political, and regional stress the best source for practical help and information.

Among the most serious scanners of law enforcement frequencies are law enforcers themselves. In some states, highway patrol vehicles include scanners as standard equipment, and all over the country police buy handheld scanners at their own expense, and use them on duty. A California SWAT communications expert explained that budget shrinks make it necessary to become more efficient, and to do that one must be aware of what's happening even outside a beat. If a cop can get the jump on his dispatcher, all the better. So if the police find scanners useful and interesting, perhaps you will too.

Scanning police frequencies isn't difficult. You can buy a scanner, turn it on, scan for transmissions, develop your own list of interesting frequencies, and eventually learn enough to understand what's happening – though that may take months of frustration and mistakes. It's simpler to seriously study how law enforcement works in your area, buy and use a frequency list, learn the lingo, and apply everything you hear to a detailed map of the city. Here's how to get started.

First, consider your objective. You wish to accurately and conveniently monitor police activity within radio range, to select the action of interest, to tailor the reception of your system so it focuses upon that area, and to correlate the radio transmissions with a street map.

To begin, you should develop a mental image of your target and an understanding of their communications practices. In most areas, local law enforcement seems to be divided into three categories: urban police, county sheriff, and state highway patrol, but it's not that simple. Local cops frequently work with federal agencies such as the Drug Enforcement Agency (DEA), Immigration and Naturalization Service (INS), FBI, the Bureau of Alcohol, Tobacco, and Firearms (BATF), and others. They also cooperate with military police and commercial security firms. Multiple agencies make it more difficult to develop a good understanding of what's going on, and to track the action, unless you are well organized.

Experienced scannists will program all the important police frequencies on one or two frequency banks, or "buttons," and then put the local supporting agencies (fire, paramedics, hospitals, etc.) on a second , federal agencies on a third, military security on a fourth, industrial security on a fifth, and so on. A 200 channel scanner has 20 channels on each of the ten number buttons, and 20 are generally sufficient for any one of these categories. See *Figure 37*.

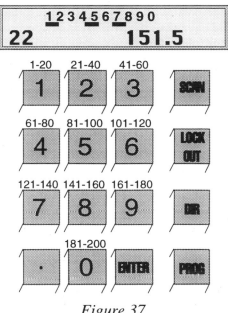

Figure 37

The keys can be divided in many ways, according to the specific situation and the interests of the hobbyist. One technique is to make key 2 NORTH, key 0 SOUTH, key 4 WEST, and key 6 EAST. Another useful method is to put the twenty busiest police channels on key 1, the next busiest block on key 2, and so forth. Most hobbyists establish one key for each of fire, highway patrol, emergency services, federal agencies, and so forth.

Communication can be divided by geography, where unique frequencies are assigned to each sector of the city. Communication is also divided by agency, with highway patrol, sheriff, and police all on separate sets of frequencies, and even specific functions within those agencies are assigned their own frequencies. Detectives in the NW segment might operate on different frequencies than those in the SW segment, and on different frequencies than motor patrol officers.

As the action evolves, additional banks of frequencies can be brought on line with a touch of the appropriate button. This allows the listener to "shape" the reception of the scanner to the situation, opening reception to the agencies that are working together. Ultimately, the hobbyist can become an expert, and that's the objective. To reach that level, begin with an overview of the police departments and the way they communicate. In medium to large metropolitan areas, and to a lesser degree smaller urban centers, the police communication system is complex and requires considerable study before it makes sense.

Regardless of the method selected, it's wise to write down the programming, cover the print with clear tape or plastic, and keep the card handy.

In a large city, if the receiver antenna is near ground level it may be impractical to expect reception from more distant segments. And remember, when you load a scanner with frequencies that aren't of practical use, you're wasting scanning power.

Throughout the country, most *conventional* law enforcement
communication is within one or more of the following FCC-
allocated bands (in MHz):

LOW:	37.02-37.42, 39.06-42.94, 44.62-46.00
HIGH:	154.625-154.965, 155.100-156.015, 156.020-156.210, 158.73-159.225
UHF:	453±, 460.000-460.550, 465.000-465.550

Searching among these bands is time consuming and
inefficient, when buying a frequency list or making a phone
call to the police communications manager will usually
produce more information than you need. By referring to
Hollins Radio Data's *Police Call*, or a similar publication, the
problem will be to identify the most useful of the hundreds of
frequencies assigned to local law enforcement agencies.

One police communication problem is density. Ordinarily,
there's a separate frequency for each area, and when it's busy
the rest of the patrols in that area must wait their turn. In an
emergency, police obviously don't want to wait for a
transmission to end before beginning their own, but they
must if there are not sufficient channels assigned to that area,
or to that type of operation.

Almost as significant a problem is efficiency of usage, since
the FCC is under great pressure (and transmits that pressure)
to ensure that the spectrum is efficiently used. Some cars
must patiently wait their turn to transmit on their assigned
channel, even in an emergency, while other frequencies are
silent 90% of the time. Something's wrong.

Trunked Communications

One solution – and perhaps the best – is called "trunking." It's
a way to use the spectrum more efficiently, to compress more
communication into fewer frequencies. Trunked systems
could have been established at any point in the spectrum, but

in fact the industry moved to a new frequency that was once too expensive for our civil servants: it's the 800 MHz band, just above cable TV, and just below cellular.

When the trunking concept was developed and then accepted by the FCC, the intention was to support dispatch services. Since then the concept has been stretched to include other radio functions, including paging, but in the eyes of the FCC dispatch is the primary purpose of the system and that function has priority in all allocations. Trunking simply makes communication more efficient, and in radio, "efficiency" is defined as how much communication can occur within a given segment of the spectrum, because – as always – spectrum is a precious resource. Without question, as communication continues to stress the spectrum, trunking will evolve into the primary technology for all dispatching.

There are three major varieties of trunking systems, developed by LTR, GE, and Motorola. They're competitive and non-compatible, and often more than one system will be used in the same area. Trunking is used by law enforcement, other emergency services, and industry, and often more than one agency will use the same frequencies.

Trunking would be impossible without computers. Shared channels are established for a large area, supported by repeaters so a car, or even a handheld walkie-talkie, can access the system from anywhere in the area of jurisdiction. If twenty channels are used, nineteen are for voice communication and the twentieth is a management channel that carries channel allocation instructions to the system.

The basic principle is the same whether a trunked system is operated by the police department as a Sole User, or by a Specialized Mobile Radio (SMR) vendor, who sells access to his system to various unrelated commercial customers. Each user group is called a "fleet" in trunking parlance. A fleet might consist of an airport shuttle service operating ten vans, or a police force with 200 users. *Figure 38* shows one arrangement whereby a bank of channels is "managed" by computer instructions on one of them. Other trunking schemes bury the queries and instructions within the data channels, and do not use a management channel.

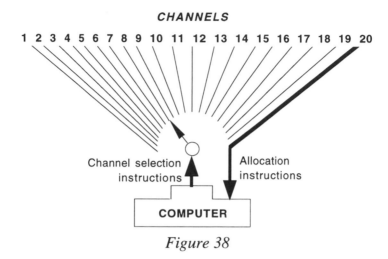

Figure 38

The principle of trunking is to establish a *bank* of voice channels that is available to *every* user (instead of using one or two frequencies in each area), and then use a management channel to assign an open frequency to each and every transmission. When a transmission begins, the computer in the transmitting radio first "asks" the main computer which frequency to use, via the management channel, and then shifts the transmitting radio to that frequency. That happens in a fraction of a second, even before speech begins. In most systems, the transmitter and the receiver remain on the same frequency until the communication sequence is complete, but in others they hop around with every click of the transmit button.

Complicated for us mortals, but a trivial task for a computer, and remember, the request for an open channel, and the response, both travel at the speed of light. Only the computing takes significant time, and computers can be very, very fast.

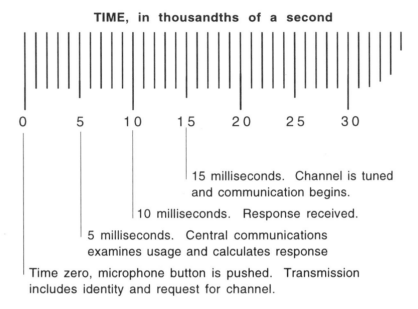

Figure 39

In considering *Figure 39*, assume the user wishes to transmit the phrase "STATION ALPHA, CHECK IN." By the time the "S" in the first word is *half-uttered*, the communication will commence. The computer protocol and digital dialog take less than one fiftieth of a second. You won't miss it.

When multiple public agencies share a single set of trunked frequencies, it is possible that the system will occasionally become saturated. In such cases, the computer assigns priorities to each agency and manages calls accordingly. Rescue and medical units are given the highest priority, the police get the next, and the tax collector is last, of course.

What happens when a trunked system is saturated, and a user encounters an emergency that is truly critical? Trunked radios are usually equipped with an "EMERGENCY" button, which cuts off the user with the lowest priority and assigns that channel to the emergency.

Some trunked systems shift the management channel periodically, even daily. Because it is nearly constantly transmitting digital data, it will stop all scanners without "smart squelch." The management channel is filled with the meaningless digital jumble of sounds. Can a scanner decode that information and shift automatically with the police radio? Not yet, and perhaps never. To do that might require a computer, plus access to the algorithm (program) used by the trunking equipment. While the frequencies are in the public domain, such software is not. On the other hand, there *are* ways to listen to trunked transmissions.

Trunking systems are assigned frequencies somewhere between 815.000 and 868.9875 MHz. Smaller systems use five channels, and until the assigned space becomes crowded, the channels are separated by 1 MHz. Frequency books sometimes show the trunked band as "855-861," which means there is trunked activity at 855, 856, 857, 858, 859, 860, and 861 MHz.

Using portable radios that depend upon a distant computer is useful for security reasons, also. Suppose a handheld radio is stolen or lost? When it next comes on the air, it must identify itself and await instructions. If the central communications computer knows that unit has been stolen, it can send a silent digital instruction that makes it brain dead and useless to the thief. That word will propagate quickly through those who might be interested in stealing such communications equipment, and the theft rate will fall accordingly.

So trunking is important because it permits more efficient utilization of the spectrum, enhances security of communication, and makes stolen radios useless. Trunking is here to stay, to the displeasure of the scannist.

Considering current technology, the best way to monitor trunked communications is to sit next to the dispatcher and listen to the loudspeaker. Second best would be to pull the plug on the central computer, or to transmit erroneous data that would "not compute," since the "fail-soft" feature of some trunked systems will shift the system back to conventional non-trunked radio. You can listen till they kick down your door.

If your police department is touchy about things like that, use a fast-hopping scanner (15 channels per second or better) and program each of the trunked frequencies into its own channel. For instance, if there are 18 such channels plus one management channel, they should all be in the same "bank" in scanners so equipped. Turn off DELAY (channels can change during pauses). Scan, and when the system stops on the digital hash of the management data, *LOCK-OUT* that channel. Then, keep your finger near the *SCAN* button. When you hear a transmission you wish to follow, hit the *SCAN* button when the unit stops on anything other than that conversation. It's not easy, and it's often chancy, but it works. It works best during the day when radio density is light, but even at night, in those areas where everything is trunked, it's the *only* thing that works.

There is a low probability that scanners will be developed with automatic trunking capabilities, for two good reasons. For a radio (scanner or otherwise) to follow trunked calls requires some implementation of, or adaptation to, proprietary system software. Motorola, LTR, and GE all market their systems with "increased" security as one feature, and are unlikely to license scanner makers to install or use their copyrighted software. To protect their proprietary positions, therefore, a lawsuit is likely if a scanner were to violate that copyright even though the scanner itself (as a non-transmitter) did not compete for trunked radio business.

The second reason is that it is illegal to produce a scanner, or an accessory for a scanner, that can decrypt digitally encoded information. It can be argued that the digital information that manages a trunking system is subject to that limit, and no one in the scanner business wants their product "de-certified." For those two reasons, trunk-tracking probably will not be a feature of the next generation of scanners.

The future of industrial and public safety radio, particularly in support of the FCC-blessed dispatch function, is trunking in the 800 MHz range, and many firms are gearing up for that expanding market. New technologies, applications, and products, will continue to emerge as the trunk business grows. Unfortunately for the scannist, the device likeliest to give him access to trunked communications is his thumb upon the scan button.

Communication Between Computers

Most major metropolitan areas have found it efficient to establish background digital communication between terminals in the communication center and computer-driven displays in the police car. That capability involves a major capital investment (in the equipment) but it speeds up communications and optimizes spectrum usage, because a digital message can be transmitted in milliseconds while voice might take a minute. There are several vendors of the hardware and software used in these systems, and they all work similarly.

To begin a dialog, the officer might reach over and punch in a license plate number, and the central computer will provide a printout, with data something like:

PO   123ABC   *and then the enter key*

CTR   1986 WHITE BUICK SEDAN
RO *(registered owner)* JOHN L. SMITH
MALE CAUCASIAN, 1968 *(birthyear)*, 72", 150
POUNDS, BROWN *(hair)*, BROWN *(eyes)*, NDF *(no
distinctive features)*
1234 5TH STREET, APT 5
OAK CITY
DL *(driver license)* 12345678F.
NO WANTS OR WARRANTS
ONE MV *(moving violations in the last twelve
months)*

In addition to direct observation, the police officer collects a lot of information before he turns on his roof light, and can make a preliminary decision regarding backup, potential problems, and how to handle the driver of the car. There's a high statistical risk, nationwide, to an officer walking up to a stopped car, and the more information available to the officer, the lower the risk.

PO   *Hits the function key corresponding to "stopping the car," and the computer sends an appropriate message to the center.*

PO   *[voice]* STOP SIGN, NOW WESTBOUND ON THIRD CROSSING MAIN.

CTR   10-4

Very efficient.

The dispatcher knows the identity of the officer, the number of his car, location, the description of the car being stopped, the name and record of the registered owner *(the officer would have made a verbal correction if the car were a brown Nissan driven by an elderly woman),* and the immediate issue.

The officer knows who should be driving, what he should look like, and what record and driving history he may have. Both the dispatcher and the police officer know a lot about what's going on, but the scanner missed most of it.

That's interesting information, but rarely life-saving or critical. While the dispatcher might recall a name or connect a car description to a recent event, the dispatcher's computer has the ability to go out to a major data base, including the National Crime Information Center (NCIC) and collect even more data, when appropriate.

And that often produces results that are challenging to the officer, and fascinating to the eavesdroppers.

What about:

PO   123ABC (*enter*)

CTR   1986 WHITE BUICK SEDAN
         RO (*registered owner*) JOHN L. SMITH
         MALE CAUCASIAN, 1948 (*birthyear*), 76", 250
         POUNDS, BROWN (*hair*), BROWN (*eyes*), NDF (*no
         distinctive features*)
         1234 5TH STREET, APT 5
         OAK CITY
         DL (*driver license*) 12345678E
         FEDERAL AND ALABAMA WARRANTS, HIGH
         RISK, ASSUME ARMED
         NO MV (*moving violations in the last twelve months*)

CTR   YOUR 20? (*10-20, or "what's your location?"*)

PO   WESTBOUND ON 3RD STREET CROSSING
         MAIN. I'M A BLOCK BEHIND. NO RABBIT (*the
         objective is not yet trying to evade*). MALE
         PASSENGER. BACKUP?

CTR   FROM THE WEST, CARS 123 AND 124, ETA
         (*estimated time of arrival*) ONE MINUTE.

Computer communication does even more. Many police cars
are equipped with a system known as Automatic Vehicle
Location (AVL), which can be based on several principles.
LORAN was once a nautical navigation system and capable
of a few hundred yard accuracy. GPS (Global Positioning
System) is accurate to a few hundred feet in its non-military
version. A third system uses "dead reckoning," once the bane
of the B-29 navigator, in which the onboard computer makes
location estimates based upon speed and direction of travel.

Still another deploys a small transmitter on the vehicle (or
shares an existing one) and maintains a system of receivers
throughout the area, all transmitting to the same computer via

phone lines or a radio link. Direction-finding (DF) determines the location of the transmitter with reasonable accuracy, as shown in *Figure 40*. In that drawing, there is only one location that exists both on the 250° radial from Station A *and* on the 190° radial from Station B. Thanks to computers, the vehicle's location is easily calculated even as it moves throughout the city.

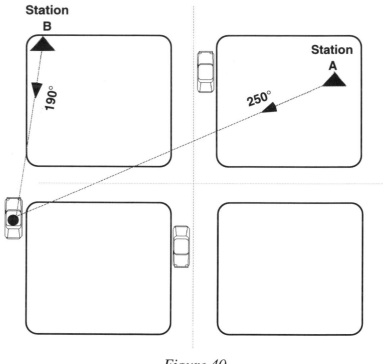

Figure 40

When a vehicle's computer transmits its automatic location information or, in the case of direction-finder systems, the dispatcher's computer transmits location data to the vehicle, radio transmissions are further reduced.

Individual officers in most jurisdictions carry personal communicators. Eventually, such units will transmit periodic silent radio "beeps" to receivers, so location will always be known, using direction finding. Further, a "man-down" feature is nearly always incorporated in modern equipment, to send a special emergency signal if the unit doesn't move for a set period of time, if the unit goes horizontal for too long, or if an emergency button is pushed by the officer carrying the unit.

Digital communication is becoming more and more common, with computers conducting most of the dialog. The scanner hears only the tip of the communication iceberg, and the rest remains a mystery. Unless... Can a scanner be configured to receive and decode computer transmissions between a patrol car and a communication center? Yes, but that's another story and it's not limited to the police.

Voice Codes

Police communication contains a lot of voice code, not necessarily to hide anything, but to shrink the dialog and to remove some of the ambiguity and negative implications of verbal description. It's both faster, and easier on the mental state of the participants in the dialog, to hear "a 187 by a 918 ADAM in a 37" than "a homicide by an escaped mental patient in a stolen vehicle."

Since secrecy is not the objective of radio codes, one would expect that the same code system would be used in all police departments, nationwide. The fact that radio code systems continue to divergently evolve might be a key (seriously, folks) to the evolution of language. There's little reason for it, but it happens as though there were some Darwinian necessity to personalize communication through innovative changes.

Back to the police... A "949" in one city might mean "gasoline spill," and in some other city might mean "hot wires down." While there is a lot of diversity among the nation's radio code systems, there is also a commonality.

APCO (The Associated Public-Safety Communications Officers) has developed a set of "10-codes" that are used by most police agencies today. Those codes, plus the current phonetic alphabet, are listed in Appendix 1.

In many parts of the country, operational function and police rank may also be expressed in code. Codes and their usage vary considerably from area to area, but as examples, radio transmissions might include something like the following:

THIS IS 22-400
(the 22 indicates the precinct and -400 means "detective")

GET ME -000
(000 usually means the chief, or the senior duty officer)

I NEED A RELIEF OUT HERE, I'M -350
(identity, and denotes day shift)

IT'S A 900
(the communication is via a handheld unit)

Some departments find it convenient to reduce shift information, rank, precinct, type of radio, events, and other information to numbers. Need a mechanic to repair a broken down, but probably repairable, patrol vehicle? Call for a "750 MIKE." Had a serious traffic accident and need an investigator? Say "11-82." Again, this communication technique is designed not to hide information but to reduce on-air time, and local frequency lists usually include glossaries.

Virtually all municipalities and area agencies share one or more common mutual aid frequencies. Usually referred to by color code ("GO TO CHANNEL WHITE"), these channels are subject to inter-agency dialog conventions and are usually free of coded references. The situations justifying the use of mutual aid communications are exactly the sort that most interests the scannist, so it's useful to know these frequencies and to have them programmed.

Mutual aid frequencies and dialog conventions are shared by police, sheriff, fire department, ambulance services, and other agencies that find it useful to work together from time to time. In cities that host a military base, the military's police and fire departments are also linked on common frequencies and share both problems and terminology.

Federal Trunking (credit to the All Ohio Scanner Club)

Some federal agencies have adopted trunking protocols. In the chart below, Wright Patterson AFB has used Group 1; other utilization is not yet known, but the assignment is published so check your recent frequency allocation books.

Group 1	Group 2	Group 3	Group 4
406.35/415/5	406.75/414.75	506.55/415.35	406.95/414.95
407.15/415.95	407.55/415.55	407.35/416.15	407.75/415.75
407.95/416.75	408.35/416.35	408.15/416.95	408.55/416.55
408.75/417.55	409.95/417.15	408.95/417.75	409.35/417.35
409.55/418.35	409.95/417.95	409.75/418.55	410.15/418.15

Police Repeaters

Public safety agencies, including the police, use two types of repeaters. A geo-fixed repeater is stationary atop a building, tower, or hill and repeats all signals it receives, usually on a second frequency separated from the receive frequency in accordance with agreements. In a trunked system, such repeaters are often intelligent or are intelligently managed by a central computer. They shift frequencies in accordance with trunking principles, awaiting channel assignments before transmitting. These systems are no tougher to monitor than non-repeated trunked transmissions.

Mobile repeaters are a different story. Many police, sheriff, and highway patrol officers carry handheld radios with relatively low power output, and use vehicles equipped with repeaters that receive such low power signals and retransmit them with much more power, on a new frequency.

Those vehicular repeaters created an opportunity that entrepreneurs have not missed. Even when the police vehicle is moving, transmissions received on certain frequencies are retransmitted on the mobile repeater frequency. That creates a network of radio energy that permits even low power radios to work anywhere and still be received everywhere. It is now possible to purchase receivers, designed to be mounted in a vehicle, that can detect those re-transmissions and thereby alert the driver to a nearby law enforcement vehicle.

It's perfectly legal to sell that hardware, and some products (Trident, for example) include other exotic features such as laser and radar detection, CB monitoring, and scanning of published law enforcement frequencies – all simultaneously.

Only in America!

And the law? Federal law makes it a crime to use any information heard *(through scanning)* to aid in committing a crime. It's also a crime to listen to and decode any scrambled messages, and since the police computers use a digital code that is intrinsically a form of scrambling, technology provides more security to police operations than our legislators have.

Though voice communication in law enforcement is shrinking, and trunking makes it more difficult to follow a conversation, listening to these communications remains the single most popular objective of the scanning hobbyist.

And it pays off.

In the event of a natural disaster or a city-wide riot, wouldn't you like to have a scanner and be able to listen to police conversations? So would the hundreds of thousands of enthusiasts like you, and while that's not a very powerful lobbying group likely to be cultivated by politicians, there's a strong constitutional argument in their favor.

There is a high likelihood that disaster control and civil defense channels will remain accessible to the scannist.

FINDING FREQUENCIES

17

Primary Resource: The Frequency List

In addition to copying a list maintained by a more experienced acquaintance at a local club meeting, the scannist has at least four resources available to help identify the "best" or most interesting frequencies in his area.

The first, of course, is the scanner itself. Through systematic scanning, listening, and note-taking, the listener can eventually develop a reasonably complete picture of the communication in the local area. In a modern urban area, "eventually" can mean years. In fact, RCMA members are constantly reporting new information to each other, and there is never a period of stability. Therefore, attempting to develop a frequency plan using time and a scanner, and nothing else, is impractical unless one's standards are low.

The second approach is to collect FCC information on frequency allocations by simply asking for it. Such allocations are in the public domain, and are freely available to the citizen. Of course, the next step is to comb through a huge quantity of information in the attempt to identify the optimum frequency list for a given area. That takes less time than starting from scratch, but it still takes a lot of time and it involves trial and error because the *allocation* of a given frequency to an applicant does not necessarily mean that the assignee will actually use that frequency. Nevertheless, it is useful to acquire the FCC's materials because as one's enthusiasm for the hobby grows, so does interest in a competent library.

A third approach is to buy a frequency book from a conscientious publisher. The best known is sold by Radio

Shack (with their imprint) and other retailers, entitled *Police Call*. It has been published for over thirty years by Hollins Radio Data, a division of Mobile Radio Enterprises.

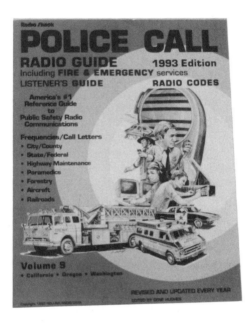

Hollins Radio Data's *Police Call* – the most widely distributed frequency list in the country.

The *Original Police Call* is produced as nine volumes for nine different geographic regions of the country. Each book is cross-referenced to include the following:

Name of licensee	FCC call sign
Transmitter city	Type of station
Frequency	Number of units

Specific assignments (often)

These manuals, with fine print on thin paper, are quite thick and contain a tremendous volume of information. It's far less

work to use *Police Call* than it is the raw data from the FCC,
and each book is certainly worth its typical price of $10,
though national coverage costs about $90.

Whether sold by Radio Shack or other retailers, *Police Call*
includes frequencies for air operations, and government
agencies, as well as law enforcement agencies. In addition,
the book uses Hollins' exclusive Consolidated Frequency List
structure, which indicates the primary type of use of
frequencies from 25 to 940 MHz.

Another excellent resource is the *Betty Bearcat Scanner
Guide*, published in seven regional volumes to cover the
United States. There are many other frequency books and
lists available, such as *Monitor America,* by Scanner Master.
Some regional publications are commercial (*Scan Fan* is a
good southern California example) and others are published
by clubs.

Even locally-published frequency books can be quite
complete if they have been published for at least a few years,
as readers frequently submit changes and corrections. There
are also specialized books, on railroads, aviation, government
agencies, etc.

In addition to frequencies, the typical book will include
geographic information, local call signs, the most-used "Ten-
Codes" for the area, the local preference for radio codes and
signals, and much more information than one would expect
for the typical $10 selling price.

Counters

There's another approach that's a bit more direct. Many
companies manufacture inexpensive ($150 or so for a basic
unit) handheld instruments called "frequency counters."
Accurate to within one part per million or better, such
instruments can determine the frequency of a transmission
within a fraction of a second. Simply hold one within a few

feet of a radio transmitter, push the transmit button, and read
the frequency on the counter. Even when users are unwilling
to cooperate with you and your new counter, there's a way. If
you can get your counter near the transmitter when it's on the
air, you can easily capture the transmit frequency, because
even a relatively weak transmitter within a few yards is likely
to be seen by the counter as a stronger signal than high power
signals from more distant transmitters. See *Figure 41*.

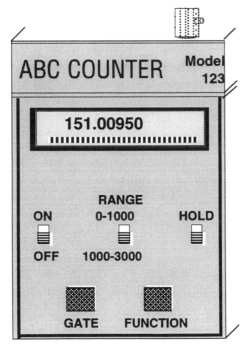

Figure 41

Search Services

Some specialized services will do "frequency searches" for a
specific geographic area. While Hollins Radio's *Police Call*
covers a general area, it includes much more than the average
hobbyist needs for his specific neighborhood. Frequency
search services use data bases developed from FCC
information, and are very focused. They're also expensive.

Some specialized services focus on spectrum users rather than geographic areas. An example is Mobile Radio Resources of San Jose, California, which publishes a list of government agency radio systems on floppy disc, in ASCII format. Such ASCII files are useful by both DOS-based computers and those Macintosh computers equipped with superdrives, and can be used with computer-assisted scanning systems such as the CommTronics HB-232.

Use Your Imagination

There's an old problem posed to high school physics classes: "How would you use a barometer to find the height of a building?" While some students talked about the adiabatic lapse rate and pressure per foot of altitude, the near-winners said they'd drop the instrument and measure the time it took to reach the concrete below, and then calculate.

The answer that applies best to the scanner hobby is, "I'd ask the building superintendent how tall the building is, and if he doesn't respond I'll say, 'Answer or I'll hit you with this barometer.'" When you want to know something, ask. The communication control/management functionaries in the public safety business are well aware that their frequencies are published and in the public domain, so just ask. They'll usually tell, and you don't have to do anything as drastic as sit in front of your scanner for hour upon hour, waiting for someone in the local bus company to say something.

But some of the information you'll get only the hard way. You may invest a lot of time and effort to find transmissions of interest, or to refine your own list of frequencies from the multitude presented in a published list. Development of a personally satisfying frequency list takes a significant commitment of time compared to some hobbies.

The return is worth it.

Of course, if you become tired of conventional books, and seeking your own solutions has become boring, you might consider one of the most authoritative frequency lists on the market, by Bob Grove (check the Bibliography). Issued as a disk for each state it misses nothing, as the entire database contains five gigabytes (that's 1/200th of a terabyte) of information.

Not even Grove is a perfect solution, however, and a lot of radio traffic occurs on temporary frequency allocations, or on "borrowed" equipment. For any geographical area, the blanks are best filled in by scanning.

Most dedicated scannists set aside a few minutes each day to examine some specific segment of the spectrum. Of course, it's easier with computer assistance to record "hits," but a notebook works just fine. "Discoveries" are welcome at the Radio Communication Monitoring Association (RCMA), by *Monitoring Times*, and local club newsletters.

THE LAW   18

This information is provided for educational and entertainment purposes, and is neither intended nor offered as legal expertise. Any reader who really needs legal advice should consult with his barber like everyone else does.

Privacy

Privacy is a key ingredient of our society, hence it is protected by laws that stem from our constitution. The sword cuts both ways, however, and we are just as reluctant to permit violations of the privacy of a citizen whether it is by society's peacekeepers or another citizen. We have generated a lot of legislation to control just who can eavesdrop, and under what conditions. Both California and New York have made it illegal to monitor cordless telephones, and other states are establishing similar legislation. Also, both state and federal laws ban the decryption and listening to scrambled communications. Collectively, however, those laws are difficult to enforce. Though virtually all receivers radiate *something*, a scanner does not radiate any identifiable signal when it's listening to cellular or cordless telephones, so there is no clear path to the evidence that will support a conviction. In fact, despite the wealth of legislation on the books, to get a conviction generally takes the cooperation of the accused.

Communications Act of 1934

It's the government's intent, and its responsibility to society, to regulate the spectrum, which is acknowledged to be a resource of steadily increasing value. Various mechanisms have been used, beginning with the Communications Act of 1934, which was a rather broad treatment of the situation that has been amended as recently as 1989. This Act is generally viewed as a useful tool in establishing regulation, and in setting protocols, but is rarely sufficient to support prosecution on matters involving privacy. That takes more laws, of course.

The "Omnibus Act"

It was actually called the Omnibus Crime Control and Safe Streets Act of 1968. Overwhelmingly approved by Congress, it was initially proposed as a weapon against organized crime. It allowed appropriate government agencies to "intercept wire and oral communications under specified conditions," and the flip side was a prohibition against anyone else doing the same things. Here are a few pertinent excerpts from the Act:

(1) Except as otherwise specifically provided in this chapter any person who:
    (a) willfully intercepts ... any wire or oral communication;
    (b) willfully uses ... any electronic, mechanical, or other device to intercept any oral communication when
        (i) such device is affixed to, or otherwise transmits a signal through, a wire, cable or other like connection used in wire communication; or
        (ii) such device transmits communications by radio...
... shall be fined not more than $10,000.00 or imprisoned not more than five years, or both.

There's much, much more in Omnibus, which is as interesting to read as the phone book (unless you're an attorney, in which case the plot is fascinating).

It implies or defines prohibitions against several forms of eavesdropping, and Section 2512 appears to ban the importation ("through the mail, or sends or carries...) of any device designed to surreptitiously intercept wire or oral communications." A quick scan through the classified section of many magazines will prove that this portion of Omnibus is at least sometimes ignored, as there's an incredible variety of devices imported from the Pacific Rim, and nothing appears to be done about it. In fact, the wording of Omnibus makes it applicable to even the simplest scanner.

In most cases, however, Omnibus is becoming obsolete, as evidenced by its successor and by the degree to which prosecutors seek any other authority.

ECPA

The Electronic Communications Privacy Act of 1986, Title 18 US Code (Public Law 99-508) is among the latest federal attempts at controlling eavesdropping and surveillance. Since 1986, it has made it illegal to monitor cellular telephone calls without a court order. It is the ECPA that is generally quoted by cellular salesmen, despite the fact that it is difficult to enforce. But, it *is* law.

As law, it's subject to change. The U. S. Supreme Court has reversed parts of the ECPA, by ruling that no reasonable expectation of privacy exists on cordless phone frequencies, which leaves *that* issue up to the states...

Federal Summary

During the last five years of legality, more than 30 cellular-capable scanners have been produced by at least four companies and marketed by a multitude of retailers. Several companies have produced and marketed downconverters to permit cellular scanning with scanners that do not cover the cellular frequencies

Those hardware *businesses* that overtly support cellular scanning have changed or closed since 1993. However...

There is no obstacle in the path of a hobbyist who builds his own cellular-capable scanner, or converter. Therefore, the industry will shift to an array of "kits," as it would be un-American to allow this marketing opportunity slip by.

Setting the Example

California is a reasonable cross-section of America, with a gross state product that would put it at about seventh in the world, if it were a nation, and uncommon ethnic and sociological diversity.

Despite the acknowledged difficulty in finding and convicting violators, several states have passed legislation designed in one way or another to limit the use of scanners to violate the privacy of others, or to endanger the public. California has been among the more enlightened in this respect, and therefore this book focuses upon that state's penal code.

From a Federal Constitutional viewpoint, California is generally considered about a five on a liberalism scale that runs from zero (Attila the Hun) through ten (Bill Clinton's grandmother), though there are local and vocal pockets of extremism at either end.

The reader is cautioned not to conclude that the California approach applies equally to his state – it simply isn't so and might not be even close, particularly in those states where electricity is a novelty. On the other hand, they can't all be quoted, and California is a pretty nice place, so California it is.

The material below is *excerpted* from the 1993 California Penal Code. The complete section is not given as that might double the size of this book, and this is already about as much as you want to know. Bracketed comments following some entries reflect the author's opinion, or comments by scannists who have (usually unknowingly) contributed.

Typical State Legislation (California)

§632.5 Intercepting or Receiving Cellular Radio Telephone Communication

(a)   Every person who, maliciously and without the consent of all parties to the communication, intercepts, receives, or assists in intercepting or receiving a communication transmitted between cellular radio telephones or between any cellular radio telephone and a landline telephone shall be punished by a fine not exceeding two thousand five hundred dollars ($2,500), by imprisonment in the county jail not exceeding one year or in the state prison, or by both that fine and imprisonment. If the person has been previously convicted of a violation of this section of Sec. 631, 632, or 636, the person shall be punished by a fine not exceeding ten thousand dollars ($10,000), by imprisonment in the county jail not exceeding one year or in the state prison, or by both that fine and imprisonment.

(b)   In the following instances, this section shall not apply:

(1)   To any public utility engaged in the business of providing communication services and facilities, or to the officers, employees, or agents thereof, where the acts otherwise prohibited are for the purpose of construction, maintenance, conduct, or operation of the services and facilities of the public utility.

(2)   To the use of any instrument, equipment, facility, or service furnished and used pursuant to the tariffs of the public utility. *[Can you interpret this as meaning that the section does not apply to the use of a cellular phone that is furnished by the phone company?]*

(3)   To any telephonic communication system used for communication exclusively within a state, county, city and county, or city correctional facility. *[You're apparently free to use a scanner to listen to cellular communications within the correctional system. If you're jailed for violating one of these statutes, this lets you stay in practice till your time is up.]*

(4)   As used in this section and Sec. 635, "cellular radio telephone" means a wireless telephone authorized by the Federal Communications Commission to operate in the frequency bandwidth reserved for cellular radio telephones.

§632.6   Eavesdropping on Cordless Telephone Communications

(a)   Every person who, maliciously and without the consent of all parties to the communication, intercepts, receives, or assists in intercepting or receiving a communication transmitted between cordless telephones as defined in subdivision (c), between any cordless telephone and a landline telephone, or between a cordless telephone and a cellular telephone shall be punished by a fine not exceeding two thousand five hundred dollars ($2,500), by imprisonment in the county jail not exceeding one year, or in the state prison, or by both that fine and imprisonment. If the person has been convicted previously... the person shall be punished by a fine not exceeding ten thousand collars ($10,000), or by imprisonment...

§633   Law Enforcement Officers – Limited Exemption From Prohibition Against Overhearing or Recording Communications

Nothing in Sec. 631, 632, 632.5, 632.6, or 632.7 prohibits the Attorney General, any district attorney,... any... police officer... any sheriff, undersheriff or deputy sheriff regularly employed... from overhearing or recording any communication which they could lawfully overhear or record prior to the effective date of this chapter.

§635   Manufacturing or Selling Devices Intended for Eavesdropping or Interception of Radio Telephone Communications

(a)   Every person who manufactures, assembles, sells, offers for sale, advertises for sale, possesses, transports, imports, or furnishes to another any device which is primarily or exclusively designed or intended for eavesdropping upon the communication of another, or any device which is primarily or exclusively designed or intended for the unauthorized interception or reception of communications between cellular radio telephones or between a cellular radio telephone and a landline telephone in violation of Sec. 632.5, or communications between cordless telephones... in violation of Sec. 632.6, shall be punished...

[The ambiguity of the first clause ("Every... another") could be interpreted as applying to retail sales of **any** *scanner.]*

§636.5 Interception and Divulgence of Police Radio Communication

Any person not authorized by the sender, who intercepts any police radio service communication, by use of a scanner or other means, for the purpose of using that communication to assist in the commission of a criminal offense... is guilty of a misdemeanor. *[If you make your living robbing banks, you might accept the threat of a misdemeanor if it helps you avoid a felony conviction.]*

§637 Wrongful Disclosure of Telegraphic or Telephonic Communication

Every person not a party to a telegraphic or telephonic communication who willfully discloses the contents of a... message... addressed to another person, without the permission of such person, unless directed to do so by the lawful order of a court, is punishable by imprisonment in the state prison, or in the county jail not exceeding one year, or by fine not exceeding five thousand dollars ($5,000), or by both fine and imprisonment.

§653t Interfering With Citizen's Band Radio Channel

(a) A person commits a public offense if the person knowingly and maliciously interrupts, disrupts, impedes, or

otherwise interferes with the transmission of a
communication over a citizen's band radio channel, the
purpose of which communication is to inform or inquire
about an emergency.

*[This could be interpreted to mean that the CB'er has rights
only in an emergency.]*

Other Legal Issues

Some states have passed legislation that restrict the use of a
scanner in a moving vehicle, or make it illegal to possess one
in a vehicle. A typical such law might read as follows:

Any person other than employees of public safety
agencies who shall operate a motorized vehicle and have
in his possession a radio capable of receiving frequencies
ordinarily used by law enforcement agencies, shall be
punished by a fine of not more than one thousand dollars
($1,000), or by imprisonment in the county jail not
exceeding one year, or by both fine and imprisonment. If
the person has been convicted previously, the person shall
be punished by a fine not exceeding five thousand dollars
($5,000), by imprisonment not exceeding one year, or by
both fine and imprisonment.

Municipalities can pass their own legislation independent of
the state, and unless it is challenged based on a conflict with
state or federal constitution, it can remain on the books and a
threat to the transient tourist as well as to the bank robber
making a getaway with the help of a scanner. As one
example, *Monitoring Times* reported that Livonia, Michigan,
passed a law that makes it illegal to have a scanner in a
vehicle. The penalty can be severe: 90 days in jail plus a
$500 fine.

What about the scannist on vacation? A typical motortrip
might pass through a dozen states and make the hobbyist
subject to a dozen different sets of legislation. The scanner

can be left at home or packed away in those states prohibiting it, or such laws can be ignored and the threat of penalty (usually a misdemeanor) accepted.

When having a scanner seems like a good idea, as it should when traveling, the experienced hobbyist will check with a local club, or review back issues of some of the periodicals that discuss scanning legislation, and lay out the route accordingly. RCMA's *Scanner Journal*, or Norm Schrein's *National Scanning Report* are good resources for travelers.

Canada's government is interested. The Communications Minister has identified cellular phone monitoring as a serious problem. New legislation is pending as of this writing, and it includes penalties of up to 1 year imprisonment and a $25k fine. In Canada, companies or media that make use of such illegally obtained information can be fined up to $75k. Those may be Canadian dollars, but it's still expensive.

Industry Concerns and Objectives

The laws of most states, and of the federal government, have been tested and found reasonable and fair by our court system, though many are difficult to enforce. One of the problems of our society is that laws that are not enforceable dilute the respect shown to those that are. Nevertheless, the consumer electronics industry and the various telephone and wireless communication industries are all coherent, and their lobbying groups are quite powerful. The goal of such lobbying is *not* a set of enforceable laws, though that would certainly be nice if it were somehow possible. The actual goal is a sufficient *appearance* of communication security as to allay the concerns of purchasers of equipment and users of services.

At the 1992 and 1993 annual meetings of the Cellular Telephone Industry Association, there were many companies exhibiting systems and software to reduce billing fraud, and many papers presented on that security topic, but no one said a word about eavesdropping. That's because it's not in the interest of the cellular industry to inform consumers that their conversations may be monitored by strangers. After all, the equipment does just what the consumer expects... and more!

We warn smokers, drinkers, and drivers of a variety of product-associated hazards, and all foods carry labels regarding content and usage. What's the chance that we'll see legislation that makes the following label (*Figure 42*) mandatory for cellular and cordless telephones?

WARNING!

Conversations on this device are not necessarily private, and can be overheard by anyone equipped with a scanner.

If the industry's true goal were to protect private information, such a label would surely be more effective than the legislation described above, yet there is virtually zero likelihood that such labels will be adopted by the industries involved. Further, some companies make or sell *both* cordless/cellular equipment *and* scanners.

The implication is that neither communication security nor the criminalization of monitoring are as important to the industry lobbyists as is the confidence of the consumer, however ill-placed that confidence may be.

Capturing Cellular Identities

Regarding the illegal use of scanners, the cellular industry's *primary* concern may be loss of revenue through fraud.

There's federal legislation that makes it a crime to alter the identity of cellular phones, because that identity is used by the system to correlate a call with the party to be billed for it.

A scanning receiver can be used to intercept and record the Electronic Serial Number (ESN) and other coded data transmitted when a cellular call is placed. Such identifying information is embedded within the first moments of communication, which a scanner can monitor. That data is captured, decoded, stored, and then loaded into a computer. Through a memory-writing device, the computer "imprints" that identity on a memory device in another phone, which is then used to make calls.

With the understanding that some cellular calls are monitored by legitimate law-enforcement agencies, people making calls that they'd rather not be traced to a person (drug dealers, for instance) might use such cloned phones.

Some such modified phones are taken to parking lots or street corners and rented by the call or by the hour to immigrants, who make thousands of dollars worth of calls to family or friends overseas. It might take many hours for the cellular carrier's computer to highlight the unusual activity, allowing staff to uncover the fraud, and meanwhile the bill is skyrocketing. Such bills are paid by the owner of the original phone if he doesn't check his invoice, or by the cellular carrier if he does, and complains. The response of the cellular carrier is to simply "shut down" the identity of the cloned phone, incurring the anger of the legitimate customer who then must have his phone legally modified, or buy a new one.

Federal penalty for cloning and certain other cellular crime is serious: the maximum is $50,000 *and* 15 years in prison.

Convictions and Decisions

Virtually all published convictions under the cellular monitoring and cordless phone monitoring statutes have resulted from either the admission of guilt by the listener, or from a proved attempt to either blackmail or sell the information thus collected. The hobbyist is unlikely to be convicted unless something active and visible is done using the information overheard. That situation may not last for long. The day may come when a scanner that has been modified *or* programmed to receive cellular or cordless telephone calls is considered sufficient evidence to convict.

The problem may not be a judge's opinion, but that of a jury. A civil suit does not require the level of "proof" required for a criminal conviction. Regardless of the primary objective of a civil suit, if a jury can be convinced that a scanner was illegally used to monitor private communications the decision might be influenced, or the resulting award might grow.

Some jurors might decide that a scanner that was found to be programmed to receive cellular telephone calls, was *used* for that purpose. Supportive evidence might include purchased information on how to receive such signals, and an antenna that is tailored to assist in illegal operation of a scanner.

There is therefore a real threat from law enforcement agencies if anti-scanner laws are broken and use is made of the resulting information, and a potentially greater danger from civil suits where a jury's opinion may help shape new laws and penalties. On the other hand, there appears only slight risk to the scannist who sits at his desk, tunes his PRO-2006 to cellular, and listens for a while. It's a pretty safe hobby unless your scanner happens to pick up your neighbor's wife making a call while he's sitting there having a beer with you.

Precautions for the Hobbyist

The first line of defense is knowledge. The hobbyist should know the laws of his state, and of the federal government, and should not break them, but under the wrong circumstances, even the most benign and law-abiding scanner enthusiast could be in serious jeopardy. Most in that category have, over the years, accumulated a variety of hardware that in the imagination of a judge or jury could be somehow damaging. Think of it! A home-made whip antenna could be measured, and if found to be exactly correct for cellular frequencies, it might be used as evidence despite the fact that it was cut for trunked frequencies a few tens of MHz away from cellular.

No scanner should be set aside while programmed to receive cellular or cordless frequencies, lest someone pick up the unit, check its programming and reach embarrassing conclusions. No hobbyist should accumulate literature, antennas, gadgets, software, and similar materials that support, or could reasonably be construed as supporting, the monitoring of cellular and/or cordless communication.

Ours is a litigious society, and the scanner hobby – like so many other vocations and avocations – creates opportunities to run up expensive attorney bills. Of course, those bills could be reduced somewhat if you followed your lawyer around for a while with your scanner on.

Precautions for Wireless Telephone Users

When using a wireless telephone, avoid using your full name, never state your address, don't access your bank account, provide credit card numbers, make credit card calls, or insert passwords. Whenever you want to chat with the Princess of Wales... visit, don't call.

And what if you believe that *you're* being monitored? There's only one thing you *can* do to get quick and serious action. Pick up your wireless telephone and call your Congressman to offer an illegally large contribution to his/her campaign fund.

Dante Alighieri
(1310)

in his commentary on
mobile radio...

**"Go right on and listen
as thou goest."**

More!

As hobbies go, scanning is easy. There's no fitness test or medical examination, no uniform, no licenses or testing, no background checks, no safety harnesses, and it's not competitive. You can read a few magazines, pick a scanner that seems to fit with what you'd like to do, follow the instructions, and allow yourself to become transformed into an island in an unending river of interesting and often useful information. It's *very* easy to get drowned by public safety agencies, our government, the military, industry, security companies, transportation, and many, many more... and for a few hundred dollars you can hear *all* of it. But what if that is somehow... not *enough?*

Some operate on the Everest principle: they push to the limit because they *can*. Others become fascinated by the scanning hobby in general, or focused upon one aspect of it in particular, and it becomes a near obsession. Extremists stay awake nights, worried that there might be a single signal in the spectrum that they have not yet penetrated – some code that remains private.

So once you've saturated the memory banks of your handheld, mobile, and tabletop scanners, amassed a library of frequency lists, erected a towering outdoor antenna, and subscribed to *everything*... what's next?

The Satellite Link

Though it's easy to find limitless telephone calls emanating from satellites 23,000 miles overhead, it's a bit tougher to find and understand the incredible array of other signals falling to earth like a perpetual rain. The basic principles are the same; an appropriate receiver is output to the antenna input of a short wave receiver, which is then tuned to the appropriate band and modulation.

There are four accessible bands in general use; that is, it's possible to acquire equipment at a reasonable cost to listen to any of bands listed below. In some cases (L and S bands, perhaps) a scanner fed by a proper antenna might work. In others, a dish and a TVRO receiver is needed. It would take a book (and it's been published, as indicated in the Bibliography) to cover all the opportunities to monitor satellite transmissions, especially when they include combinations of video, audio, and data, so this chapter will touch only very lightly on the subject.

L-band, near 1.5 GHz, on which vehicle control and position location is broadcast. Also in L-band are INMARSAT transmissions, which include data and voice. Several common scanners reach this band.

S-band, near 2.2 GHz, used by both military and commercial licensees for radiolocation, scientific, industrial, and medical applications. A very few scanners reach this range without help.

C-band, near 5 GHz. This heavily populated band is used primarily by VSATs (Very Small Aperture Terminals) , such as banks, retail stores, and other points on a star-shaped communication system. In the U. S., This usage is being taken over by less expensive fiber-optic links, but throughout the rest of the world, VSAT is a key communication medium that includes telephony, data, non-entertainment video, and more. C-band is also used by entertainment television, and for each published TV signal there are a dozen or more "hidden" signals that can be tuned. They can be reached by combining a TVRO receiver with a scanning receiver, and books on how to do it are in the Bibliography.

X-band, near 8 GHz, and in very fine steps (typically 1 kHz). Almost all military. Very difficult to tune without a lot of luck at a military surplus sale.

Ku-band, near 12 GHz. Also used by VSATs, and for the same purposes as the C-band terminals. Note that in both cases, some of the communication is via satellite (the dishes look up), and some is point-to-point (the dishes face each other). Since there is more bandwidth available in Ku-band, this segment of the spectrum represents a growth industry. Again, it takes a TVRO receiver plus your scanner to get the job done.

Most manufacturers of TVRO (TeleVision Receive Only) equipment build both C- and Ku-band equipment, and both bands are economically receivable using ordinary dish antennas. Ku requires a smaller dish for a given signal level.

One can purchase software to support tracking of non-geosynchronous satellites, including those in unusual orbits. Such software permits data acquisition, image processing, and much more. The data streaming down from such "birds" can include weather images, military intelligence such as ship locations (try decoding *that*), infra-red environmental scans and other optical imagery, and much more.

It's very busy out there...It's not practical to listen to the satellites without help, and several books and magazines are available to guide the ambitious scannist; check the Bibliography.

Nothing But the FAX

That's right. It's easy to intercept telephone calls (and by now you know that). You probably didn't know how simple it is to decode fax transmissions. Many companies sell boards that enable personal computers to do that decoding, and the result is a screen display (or printout) of faxed documents. The procedure is to (1) receive the signal, and (2) decode the fax – that's that! And just to be sure, the signal can easily be recorded and then decoded later, using available software. So faxes are more vulnerable than most believe. Sony makes a receiver (the CRF-V21) that, in addition to many other bells and whistles, automatically scans a number of satellite fax frequencies and uses its own built-in thermal printer to produce a hard copy. It costs $6,500, but for serious industrial espionage it's worth it.

There's a certain safety in numbers, though. Assume that your company is sending a confidential document by fax. It generally travels through the phone company's wire and fiber optic links, though it might go part of the distance via satellite or point-to-point microwave link. By placing a directional antenna "downwind" of a microwave link or looking up at a satellite, voice and fax data can be received, recorded, and eventually decoded.

But can a diligent "spy," or a nosy scannist, separate out your conversation or fax? At any instant, the spectrum is filled with thousands or millions of individual communication links, and while the mass can easily be intercepted it's tough to identify and find anything specific. But it is *not* impossible. Scary, huh?

Fast Food, Local Business

Most of today's fast food restaurants, and many other businesses, use low power industrial radio for order taking, inventory control, security, and other functions. From the hardware store to the sandwich counter, there's a lot going on that's scannable. Do you want to monitor this? Do you need to discover the ratio of hamburgers to cheeseburgers? You can, but there's a problem because "low power" in this case means *very* low, and to listen to the local hamburger joint requires a highly directional antenna such as a yagi, plus a good preamplifier, plus patience. On the other hand, if you really want to know what's in the "secret sauce," a scanner won't help as much as a trip to the local public library, where all the real secrets are published.

But some of us simply cannot stand to know that just a block away someone's talking on a radio... *privately.*

Software, Computers, Displays

Architecturally, a "scanner" is a package that contains a radio wired to controls, an internal computer and memory. Increase the speed and processing power of the microprocessor, and/or add memory, and the capability of the original instrument can be multiplied many times over. The basic functionality is unchanged, but assuming a decent antenna and a good location, a scanner plus an external computer is a very powerful combination.

Many advanced scanners provide means to integrate a computer with the radio, and some modification services will install an appropriate computer interface in popular scanners and scanning communications receivers. What can such an augmented scanner do? Ask the National Security Agency...

A record can be maintained of all transmissions intercepted, as to the frequency, type of modulation, duration of the transmission, and (if an appropriate recorder is used) content.

The spectrum can be constantly examined and activity displayed, using the computer screen as a spectrum analyzer. The real signal analysis is done on an internal circuit board, and the screen shows the frequency and amplitude of signals detected across the spectrum. Some such adaptations have reached remarkable performance levels. One by Ace Communications (and designed for use with their advanced scanners) covers 0 to 1300 MHz, with a 500 MHz span viewable at one time. At a very nominal charge, this accessory is competitive with very expensive complete instruments of only a few years ago.

When a conversation occurs on two frequencies, a properly equipped computer can look for new signals that arise just as one transmission stops, and can eventually make the correlation that will allow the monitoring system to listen to both sides of the discussion. It takes specialized programming to get the job done without violating the law against decryption of coded information.

The more advanced scanners are delivered with a data port by which a computer can load frequencies into memory, reducing the workload. Some blocks of frequencies can be purchased on disk, and then translated by a computer into a form usable by the scanner. Many enthusiasts feel that the most comprehensive and cost-effective frequency resource is *The Grove/FCC Database* offered by Grove Enterprises, which includes virtually every FCC licensee in the country.

A computer interface system usually comes with database software that permits the accumulation of frequencies on which communication was detected, and then selective tuning of them as desired. One of the side benefits of such systems is "negative," in a way. A list of frequencies on which spurious or "birdie" signals are detected allows them to be avoided.

Some computer systems actually control the direction of a rotating directional (high-gain) antenna, and can be programmed to develop a histogram of activity in each sector, by frequency, modulation, signal strength, and other selectable criteria. An aggressive software program (and fast enough computer) can "learn" to follow frequency-hopping signals, where privacy/secrecy is usually achieved by changing frequencies frequently in accordance with a pseudo-random algorithm.

A computer allows your scanner to generate a list of literally hundreds of thousands of active frequencies, with times of transmission, signal characteristics, and much more. The signals can be differentiated and filed according to the communication technique in use. One could spend years trying to evaluate the information collected in a single day's monitoring by a competently designed, computer-supported, broad-band scanning radio.

The CommTronics Engineering (San Diego) HB-232 Scanner Interface is a good example of available technology. It comes as a kit of parts, program disk, and a manual, and is applicable to the Radio Shack PRO-2004/5/6 and 2002 tabletop units, and also to the PRO-43/37/34 handheld products. Written for DOS-based computers, the HB-232 software also works on Macintosh computers using the SOFT-PC or other conversion software.

Products such as the HB-232 allow the scannist to seriously professionalize the scanning process, and to miss very little.

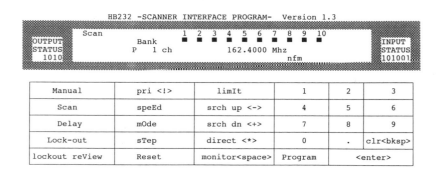

The HB-232 *MAIN* screen.

Figure 43

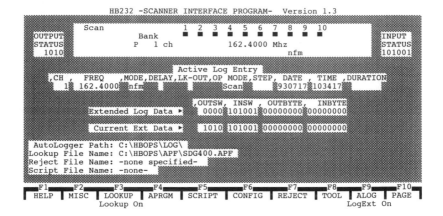

The HB-232 *AUTOLOG* screen.

Figure 44

The HB-232, and some systems like it, can do a lot to automate scanning functions. It autoprograms all supported scanners from plain text files, selectively, thus ending the tedious frequency-entry process since the software can use available third-party files and even databases generated by other scannists.

They're shared by computer bulletin-board communication (in this case, the Hertzian BBS, operated by CommTronics Engineering), or by swapping disks. It controls standard keyboard functions. It autologs information whenever a signal is detected, including such data as channel, frequency, modulation, data and time, and much more. Finally, it supports a printer, permits extensive scripting to automate otherwise manual processes, and much more.

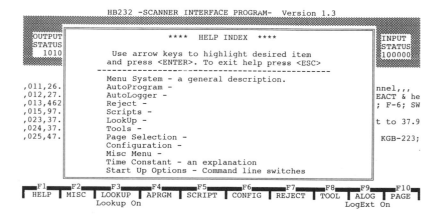

HB-232 *HELP* screen.

Figure 45

In short, competently designed interface systems allow a computer-supported scanner to become almost autonomous, and certainly much easier to use. With a good scanner augmented by such a computer and interface, the system can detect and document virtually everything happening in the æther.

With so much going on, you may never want to leave your apartment again, but if you absolutely must, technology *does* suggest an answer...

Recorders

Many extended-play tape recorders are on the market sometimes referred-to as "data loggers." Wired to the earphone jack of a scanner that's been tuned to an important frequency or band of frequencies, and set to record only when a transmission is received (signal activation), such equipment can autonomously record many, many hours of information over a very long period. Some permit accelerated playback, and use specialized tone-changing circuitry to avoid the chipmunk sound.

Such automated patience is amazing. There's no fatigue, no distraction. The machine simply sits there, tirelessly awaiting the next sound on the selected channel and then quickly switching "ON" before the first diphthong's half over. On the other hand, there *is* a disadvantage. The recorder is a time-shifting device, and doesn't replace the human, who must listen to the recorded information and make an evaluation. Nevertheless, recorders are useful, and when one is appropriate there are many inexpensive units to select from.

Decoders

Telephone touchpad decoding is a simple task, with several DTMF decoders available. Once that trivial challenge is met, equipment can be bought to decode a wide variety of other signals. There's no need to explain/define the many protocols and data mechanisms on the air, but the worried reader will be vastly relieved to learn that all of the following acronyms can be decoded or supported by one scanner/radio accessory or another, readily available on the open market.

FAX	WEFAX	ASCII
RTTY	SITOR	AMTOR
NAVTEX	AMTEX	TDM
SIAM	PACSAT	SSTV
DPSK	OSCAR	PACTOR
BAUDOT	PACKET	

Bulletin Boards

A Bulletin Board System (BBS) is a computer wired to a modem wired to the telephone system, with software to enable communication with computer/modem/software combinations that call in.

The simplest BBS permits a caller to "enter" the computer's information storage system, often using passwords, and then to deposit or retrieve information, read and leave addressed correspondence, and to read posted bulletins/notes left behind by other travelers. The more advanced BBS is a complex enterprise, whereby several telephone lines permit multiple entries, each at a password-defined level, and even interaction between callers. BBSs are established and maintained primarily for fun, usually by hobbyists to support a special interest group. Often, however, there is a validation process and fee involved before the caller will be authorized to do more than skim the surface of the system.

Many BBSs support amateur radio, short wave listeners (SWLs), and scannists. Such "boards" are often operated by local ham and scanner clubs, and provide the latest in frequency usage, commentary on legislation and FCC regulations, equipment evaluations, and other information of general interest to the hobby.

Some BBSs are more aggressive, and for validated and paid-up callers they store lists of fast-changing frequencies of sensitive communications, encryption techniques, passwords and codes (often illegally obtained), details of equipment modifications, and much more that's supplied by the membership, and sometimes the information is of questionable legality. Such boards are for those who are so focused and dedicated that they might spend many hours and a lot of money to crack an encryption scheme that they *knew in advance* protected only inventory data from a retail store. But once a coded communication has been successfully compromised, sharing the method or algorithm with other board participants earns a merit badge for the hacker.

The Hertzian BBS, in San Diego, is used by scannists and SWLs around the world. It has grown into a key resource for serious hobbyists, who share frequencies, scanning strategies, technology, and other information. Running such a BBS, according to System Operator Bill Cheek, is like being a spider at the center of a web of information. For the callers and users, on the other hand, such a BBS is a resource that is unmatched in the published literature.

With more than 900,000 members, America OnLine (AOL) is effectively one of the world's largest BBSs. It has an area in the *Lifestyles* section reserved for amateur radio and scanners. It includes discovered frequencies, user data, private advertisements, and a lot of technical help for the scannist. Other online services have similar services.

Beyond Scanners

There's a natural progression from a $80 scanner to a $500 scanner as the hobbyist seeks more performance. Once there's a state-of-the-art scanner in the pocket, in each car, on the nightstand, and on the desk, what's next?

Fortunately, the prescription for the next step is written in some of the scanner hobby magazines. Makers of short wave receivers advertise there, alongside the Unidens, knowing that as the scannist becomes more and more immersed in the technology, the equipment, and the language, capturing his next pocketful of discretion money is easy.

It's no challenge to spend ten thousand dollars for a receiver, supporting computer, a directional antenna, and a rotator, so the SWL hobby beckons hypnotically to he who has scanned it all. Like scanning, there's no physical to pass, and the only entry test is when your credit card runs through the slot. But SWL *is* a whole new world. Compared to scanning, this community has made a much larger investment, is technically more competent, spends more time at the hobby, and interchanges information more formally and actively.

If it's interesting to listen to police conversations downtown, what value should be placed on the ability to listen to communication from everywhere on the planet? If it's satisfying to have a dozen or more knobs on a scanner, how will it feel to own a receiver with more than 50 controls?

It's easy to make an expensive mistake when buying a first scanner. In SWL equipment the error can leap to thousands of dollars, so the wisest first step into SWL is probably to join a club. Some clubs (Appendix 2) welcome members from both the scanner and the SWL hobbies, and that pollination and information exchange helps reduce risk of error. The best first step into SWL is probably either used equipment (check the club newsletter or bulletin board) or a low-priced digital receiver in the $300 range. That's a temporizing measure at best, and the next step is of course to spend serious money...

And after that? There are only two places to go. If you just want to listen, you can apply for a job at the National Security Agency just off I-295 in Maryland – and an easy drive to the best seafood in the world. And when listening is not enough, the Big Transition beckons. Now that the FCC has established a simple and easy process to an amateur radio license, Morse code is no longer a barrier to the ham world. Pass a simple 55-question written test (administered by other hams), buy a handheld transceiver, and you've made the shift into the world of emitters. And then scannists can scan *you*.

Nothing's Perfect

With the right equipment and software, and with enough time, little will get past you. But even if you evaluate a transmission every ten seconds, never sleep, and have eight personalities working at once, there still isn't enough time. So prepare yourself. Recognize that no matter what you do or how much money you spend, someday, somewhere, someone will have a conversation with his gardener about pruning the roses, and you will only hear *part* of it.

Willie Sutton

when asked why he
robs banks, answered...

"Because that's where the money is."

CRIME AND WIRELESS TELEPHONES

20

Breaking the Law

For the criminally-inclined, breaking federal and state laws can be rewarding and the wireless telephone is a fine tool. The necessary tools and technology are easily found in retail stores, and whatever might be missing can be found in advertisements carried by many reputable magazines. Define the task, however illegal, and the support to achieve it can be purchased.

Only in America can one openly purchase transmitters that are intended to be used exclusively and solely to listen to wired (non-cellular) telephone conversations; they're advertised in many periodicals and sold to wannabes in spy shops across the country. Miniature "bugs" take many forms and are usually very difficult to locate and identify. In fact, one nearly invisible model is built inside a "splitter" used to couple two phone cords to one jack. The eavesdropping industry is big business and has been for many years. Wireless technology makes eavesdropping simple and virtually undetectable.

But wait... there's *more!*

Basic Principles

A scanner can be used to listen to voice communication, to touch-tone dialing of passwords and account numbers, voicemail systems, computer communication, and much more. A recorder can be used to capture the events scanned, and they can be later analyzed and, where necessary, decoded. For years it has been dangerous to process sensitive information by wireless means, and as technology improves and more information is accumulated, the danger is increased.

A DTMF decoder can convert audible touch tones to displayed numbers, thus allowing the listener to duplicate any entries overheard. That includes bank account numbers, passwords, dialed phone numbers, credit card numbers, and anything else that is likely to be communicated by touching those keys.

Some companies print a "reverse telephone directory" that is used to look up the address and subscriber information that correlates with a decoded DTMF touch tone entry. That is, if the eavesdropper hears/records a call being made, it's not impossible to determine exactly to whom it was made. It's more difficult to learn who made the call, but not impossible for a computer-augmented system. One such national directory is published on compact disk (CD) for computer use, and is claimed to contain 90 million (!!) names. Huge cast, lousy plot...

In short, the dedicated scanner-assisted crook can get a lot of help from a scanner, a computer, a DTMF decoder, and a tape recorder. These are not difficult tasks, and though in most states legislation makes such activity a crime, it's hard to convict under these laws.

Industrial Espionage

Companies spy on each other more and more, as a cost-effective means of keeping track of competition, learning about orders and business opportunities, and studying executives to determine how they'll negotiate.

The advent of the cellular telephone and cordless telephone makes the job of the industrial spy much easier. The people at a level likely to control or possess key information are often those who are routinely equipped with a cellular communication capability.

Attorneys and accountants spend a lot of time on their cellular phones because those are billable minutes. Follow the car carrying one of these professionals and he or she will be on the phone almost constantly. It's simple to use a scanner and a tape recorder to monitor and make a record of the conversation, for later analysis. To an investigator's business client, hearing his arch-rival's voice discussing strategy with his attorney is likely to be impressive.

Some people conduct a lot of business over their residential cellular phone. It takes little to park nearby (or, in a serious surveillance, set up in a line-of-sight apartment and use a directional antenna), and in some states it is quite legal to listen.

Local, Wide, and Metropolitan Area Networks (LAN, WAN, and MAN) are vulnerable because that industry is moving from wire to wireless, and the frequency range allocated by the FCC (902-928 MHz) is attainable by many scanners. It's a relatively simple matter to eavesdrop upon such frequencies, and with the right technology one could camp outside a building copying critical business data.

Generally, however, it is illegal to make use of such monitored conversations to gain an unfair advantage. Further, certain categories of industrial information are considered proprietary, and it's against the law to disseminate such data without consent of the originator.

The competent spy doesn't worry much. With an intermediate-grade scanner, a tape recorder, a directional antenna, and time, any determined investigator can find out almost anything.

Credit Cards

Any consumer who gives credit card data over a cellular or cordless phone (or, for that matter, to anyone who telemarkets that consumer) deserves the multi-pronged result. First, they'll be liable for some part of the money stolen by the criminal that listened in, and second, their long-term rates and fees will rise to compensate the card company for such crime. And finally, they'll feel like a fool. Appropriately.

Lawbreakers and Scoundrels

There is a range of violation and violators. Hobbyists who do nothing with what they hear constitute the vast majority of eavesdroppers. A few are criminals who use the technology and equipment to steal what they can, as quickly as they can. There are also a few who use scanners to carefully and occasionally augment their status or income.

Finally, there is a category of malicious people who write virus programs for computers, put graffiti on art, transmit fake distress calls, set fires, and use scanner-derived information as viciously as possible. They listen as hotel reservations are made, orders are placed, and to assignations between lovers, and simply do as much damage as they need to do in order to elevate themselves or achieve some sort of perverse satisfaction. There is no protection from those who take pleasure from injuring others, but one can reduce personal vulnerability by remembering that there is no security in wireless communications.

More Precautions

"Just because you're paranoid doesn't mean there's no one after you..." There's something to be said for paranoia, especially when radio communication is involved. Watch what you say, and never enter personal numeric data into any phone! And carefully read the new book, *Cellular Fraud* by Damien Thorn (Bibliography).

CLUBS AND ASSOCIATIONS 21

Bearcat Radio Club

Endorsed by Uniden and by DX Radio Supply (one of the central figures in scanner retailing), and with membership applications packaged with Bearcat scanners, the BRC is the country's largest scannist club with over 10,000 members.

For $30 a year, the BRC provides a one year subscription to the *National Scanning Report*, privileges on the club's Bulletin Board System (513-298-3663), access to an 800-number for technical assistance, a frequency list for the new member's county, a copy of the regionally-applicable *Betty Bearcat Scanner Guide*, and much more. The club offers two less expensive membership levels with correspondingly fewer privileges. This organization offers a lot of information and a certain camaraderie, plus local chapters and informally organized groups that periodically get together.

The *Betty Bearcat Scanner Guide* for other geographic areas can be purchased for about $8 plus freight, and it takes seven to cover the country. One of the nicest features of the club is the BBS, which provides very timely and often quite focused information on frequency changes, etc. Members are active and a lot of data changes hands, but it's sometimes difficult to get through to the BBS. Be persistent – it's worth the effort.

ANARC

The Association of North American Radio Clubs is a network of special interest groups including a variety of amateur radio specialists, short wave listeners (SWL), and scanners.

Though ANARC is not focused exclusively upon scanning, several monitoring clubs are active members.

The RCMA

One large national scannist organization (about 2,000 members) is the Radio Communications Monitoring Association (RCMA), Inc. It's the largest member of ANARC that is most interested in monitoring/scanning, and describes itself as the "largest national organization devoted to the subject of radio communications monitoring and the hobby of scanning." Whether or not that is true (the Bearcat Scanner Club has a larger membership), the RCMA is a dedicated group of hobbyists distributed throughout the nation.

From its headquarters in California (see Bibliography), this group publishes the *Scanner Journal* (formerly *RCMA Journal*) monthly. That periodical is among the most useful documents for the dedicated, serious scanner enthusiast, as it contains many articles on products and technologies, plus "how-to" tutorials by technically qualified contributors.

The RCMA tries to establish an associate editor in every state. Each is an expert in the hobby, and also sufficiently public-minded to take on the often thankless task of inputting regional information to the *Journal's* editors. It's a very aggressive and extremely competent periodical supported by a relatively small population (when compared to other hobbies and their newsletters).

And if a member believes something important is missing from the list, he/she is welcome to make a contribution and fill the vacuum. If the editorial group feels the new material is worthwhile, it might become a regular feature. Past issues of this periodical have included both limited commercial advertising and personal classified ads (for members); the latter acts like a trading post, and is reported to be an effective swap meet.

The *Scanner Journal* is organized as follows:

Aircraft	New Members List
Amateur Radio Communication	New Products Review
ANARC Report	News Media
Beginners and Experimenters	Peter Portly Adventure
Below 30 MHz	Public Safety:
Calendar	California
Canada	Great Lakes
Coastal and Inland Waterways	North Central
Commercial Communications	North East
Contact/Info Wanted	South Central
FCC Topics	South East
Federal Government/Military	West
From the Captain's Desk	Publication Announcements
International	Railroads
Low Band Skip	Shortwave Radio
Marketplace	Space Communications
Monitoring and the Law	Sports/Special Events
	Technical Topics

Everyone who uses a scanner will benefit by joining this organization because it's inexpensive ($24 per year at present), yet the advantages are many. In fact, some of the information freely published in their periodical is simply unavailable from any other source.

Other Organizations

Many other groups have formed within the scanning hobby, and they are entertaining, self-protective, and useful. Most scannists belong to at least two clubs or associations, one at the national level and another that is regional or local. Some of the organizations have developed a certain political focus, or lobbying ability, though none known to the author has a professional lobbyist working in Washington.

In addition to RCMA, which is a national organization with local chapters, one can find local clubs by visiting stores that specialize in ham radio or CB equipment (rarely will both be sold by the same shop, although both are likely to sell scanners). There will usually be a bulletin board that posts notices of the meetings of hams, SWLs (short wave listeners), and scanner/monitor enthusiasts. Also, the *Scanner Journal, Monitoring Times,* and other periodicals in the Bibliography publish monthly lists of club meetings scheduled for various geographic areas.

Some clubs are members of RCMA, others are full members of ANARC, and still others are not affiliated with any national group. In any case, these clubs have a lot of information on the local usage of the spectrum, and are a key resource to anyone interested in scanners and monitoring.

Though this hobby is usually performed solo, it is highly dependent upon collective activities and shared experiences because organized scannists multiply their information by sharing it. Whether a news flash involves a new FCC frequency allocation, proposed legislation, an interesting transmission pattern on some channel, or a disappointing piece of heavily advertised equipment, organization and information-sharing helps every enthusiast enjoy this hobby.

National, regional and local clubs are listed in *Monitoring Times* (see Bibliography) and in other scanning and amateur radio periodicals. Those listings often include membership and meeting information.

Despite rumors, scanning is *not* addictive – at least in the clinical sense. At club meetings, only rarely does a member stand, face the group, take a deep breath, and say...

"Hello. My name is Fred, and I'm a scannist."

DIALOGS 22

Perception

In the story about the elephant and the blind men, each "saw" the animal from his unique vantage point. The man who held the tail thought an elephant was much like a snake, the man feeling the leg thought it was more like a tree, and so forth.

The scanner industry is about the same. Depending upon who is speaking, it can be described as a fascinating hobby, an amazing technological achievement, a business *threat*, a business *opportunity*, a criminal activity, a test of the Constitution, a technological challenge, or a waste of time and attention.

The information below was derived from dialogs with people holding each of these opinions. Some of the discussion was recorded and is reported verbatim.

The opinions reflected here are *reported,* not *endorsed.*

Electronic Engineer

Q   ...5 kHz steps, continuous up to 1.3 GHz, and some of them do 100 channels per second. A few can do 50 Hz steps.

E   Are you crazy? That's better than ten millisecond switching speed because it has to dwell at each frequency long enough to see whether there's anything there, right? That's really a tough spec. *Really!*

Dealer/Retailer

They stand around on dark street corners, waiting for yuppies who cruise by slowly in their Beemers, looking for a deal.

"How about a scanner?" one hisses. *"Already modified. Gets cellular, and covers the military like a blanket. Only $600 with a discone antenna."*

Q   Is that a CB?

D   No, it's a scanner. See?

Q   How many of these do you sell?

D   When they first came out last fall we couldn't keep them in stock. The manager kept a list of people to call. They sold as fast as they came in. They sell slower now, but still better than, say, stereo receivers.

Q   Is it legal to listen to phone calls?

D   I think there's a rule against it, but how would anyone catch you?

Q   What about cordless phones?

D   You don't have to modify the scanner for that. Cordless is at about 50 MHz... You know, like '50 on your dial?' Almost all the scanners get that.

Q   This cordless phone says it has a 'security system.' Does that mean it's secure from scanners?

D   No. The security thing is to keep a neighbor from dialing out on your phone. A scanner can still tune it. You can listen. It's perfectly legal. We sell tons of scanners, and we wouldn't break the law. What you do with it later is your business.

Police Officer

PO   I've got one. It was damn expensive, and I paid for it out of my pocket, but I really like it.

Q Don't you get tired of law enforcement when you get home?

PO You don't understand. It's for the job. Nowadays you've got to have communication. You have to know what's going on, what's coming at you.

Q Did you always use a scanner?

PO Only the last two or three years. Look at how small this is [*holding the scanner*]. I've got a stereo, and I built speakers for it. How do they get such sound out of a tiny speaker like that?

Q I don't know. Do all of you carry scanners?

PO I do. Some of the other guys do. My brother is trying to get on the job, too. He has a couple at home, and he listens a lot. He's hooked. If he could afford it he'd have scanners hanging all over him. He really listens a lot. It's his hobby, and he's serious about it. I think he'd rather buy a new scanner than a car.

Former Sailor (*Once Assigned to NavSecGru, a Naval Subset of the NSA.*)

Q So you worked with the National Security Agency?

NSG We were in a little cell aboard ship. Only a few people knew what went on in there.

Q What did?

NSG You can guess. The NSA is a snoop.

Q Do you know what this is?

NSG [*Laughing*] Sure. It's a scanner.

Q Here's some of the spec. [*Hands over operator manual, opened to specifications*]

NSG Fast. Up to 900 MHz, right? Very interesting. It does FM, VHF, UHF. That's impressive. We could have used this. I bet it covers cellular, right?

Professional Scanner Modifier (*Phone conversation.*)

PSM   It's $290, already modified.

Q   You mean, even *after* you modify it you can sell it to me for $60 less than the store?

PSM   Absolutely, and you'll save sales tax, too. I buy them by the case. I sell lots more scanners than a store does. I'm a specialist.

Q   Will it get cellular?

PSM   I'll unblock the cellular band at the base price. Everything else is extra. Let me tell you what I can do for it.

Cellular Expert

(*Casual conversation with a cellular technologist/marketer at the Cellular Telephone Industry Association meeting in Dallas, TX.*)

Q   What about scanners?

CE   What about them? It's a problem. The subscribers don't know it, though. You wouldn't believe what goes on the air. It's like a soap opera, but with no censorship.

Q   Do you use a scanner to listen to cellular telephone conversations?

CE   [*Annoyed*] I've got other things to do.

Q   Is your industry taking precautions against scanners? What are you doing about privacy?

CE   The subscriber can do one of three things. He can watch his mouth, buy a scrambler, or wait.

Q   Wait?

CE   Sure. We're shifting to TDMA *[Time Division Multiple Access, a digital and intrinsically encrypted modulation technique]*, and that's the end of scanning.

Publisher of Frequency List

P   It's good business. Freedom of information and all that. What I do is publish information that is in the public domain. I've been doing it for a long time. It's a second job.

Q   Is it in the public's interest to give criminals the frequencies used by police?

P   That's not even in the top third of our problems. Don't be silly. How often does a crook get caught with a scanner? It just doesn't happen. Don't ask me why. If I decided to rob a bank, the first thing I'd buy is a scanner.

Q   And then?

P   I'd buy my frequency list. [*Laughs*]

Ad Salesman for Magazine

Q   You solicit advertising for services that allow the buyer to do something illegal?

M   We don't look at it that way. We don't censor. Freedom of the press, right? If we start, where would it end? That's the price you pay! We'd get crucified if we censored those ads.

Q   What if someone answered an ad in your magazine, and eventually committed a crime with the result. Would you be responsible at all?

M   That's been tested. It depends. [*Pause*] One of the books lost a big case. Did you hear about that? Is that why you asked? Did you see an ad that looks like a problem? Are you in law enforcement? We *do* screen, and when there's a question it goes to the attorney. Are you an attorney? Come on! You're an attorney, right? *Right?*

[*Editor's note: The last comment was not recorded, and was heard from a distance while the listener was moving rapidly.*]

Private Investigator

(*Over the phone, as a prospective client seeking a specific item of information on a competitor.*)

A How I do it is my business. I have my own methods and contacts.

Q But I don't want to waste my money. I want to judge whether you operate efficiently and give me my money's worth.

A My partner and I use every technique. I told you.

Q But do you listen to cellular calls?

A I said, I use every technique. Every technique. What the hell do you want?

Hobbyist (*At local ham club meeting.*)

Q What kind?

H Alinco. It's a newer brand than, say, Kenwood and Yaesu, but it's good. I think that to get a share of the market they offer more for the dollar than the established brands. Mine's been pretty well abused and it holds up well.

Q Do you use it as a scanner, too? I mean, is it any good as a scanner? Does it get police?

H Just to listen to cellular. But it's slow. On the other hand, I'd have paid the $400 just for the H/T without the scan function, so why complain? The cellular is a bonus.

Q Did you modify it?

H It just takes a minute. Look. [*removing battery pack*] Up there. See those screws? That's it. Pull them and there's a single wire to cut. Blue. Then you reset the computer control and you're on the air. It doesn't do the right steps, though. You know, 30 kHz? It misses a lot. But it's free.

Q How did you figure out how to modify it?

H   You can call the distributor and they'll send you a sheet of instructions. You don't need them, though. Any idiot can do it with instructions over the phone. You don't even need a soldering iron.

Q   Is it legal to listen to cellular calls?

H   In some states I think it is. I'm not sure about [*deleted*].

Q   I heard there was a federal law. Since 1986.

H   Maybe you're right. But who's watching?

New Scanner User (*Met at amateur radio retail store.*)

SU   It's a hobby. I'm pretty new at it.

Q   What do you listen to most?

SU   A lot of police. I've got a couple of the frequency books, and they've got mall security, government agencies, everything. It's unbelievable.

Q   How long have you had your scanner?

SU   Since the riots in LA. I figured that the best information would be what the police tell each other, right? You get uncensored information, without the radio stations putting spin on it.

Q   Why did you choose that particular one?

SU   I wanted one right away, and a friend said Radio Shack has a lot of them, and they're handy. I never noticed before, but he was right. I got an education, too. It's not quite what I wanted, but what the hell. I'll keep it for a while.

Q   Does your scanner get cellular phone calls, too?

SU   No, they're too high in frequency. It gets cordless phones loud and clear, though. I've got an antenna for that and, you know, baby monitors. Amazing. Those people never turn them off. Amazing. And it's legal, at least here.

Author's note. After a while, those who use scanners to monitor cellular or cordless conversations talk freely about it, just as those who regularly exceed the speed limit talk freely about *that.*

Alexander Pope
An Essay on Man
(1734)

"Presume not God ever to scan."

THE FUTURE OF SCANNING

Legal or Not?

There's a very good set of arguments for making scanner radios illegal under special or general conditions, and in some local jurisdictions such laws have been passed. It might happen at a state or federal level, or it might not.

Some feel that the degradation of our social fabric will ultimately lead to either a tough far-right response or to a degree of anarchy. Others believe that the liberal movement cannot be stopped, and that eventually our Constitution will be interpreted in the most liberal manner possible. Every extreme is possible, and supported by both technology and the nature of our society. It's equally possible that scanners will eventually become illegal nationwide, or built into pocket entertainment devices carried by everyone.

The only certainty is that there will be more legislation, not less. The inevitable introduction of scanner kits that allow the consumer to circumvent the FCC's limitations on manufacturers will eventually be responded to by telephone industry lobbyists who press legislators to close that gap, but another loophole will be found and the cycle will begin anew. Like the cable descrambler industry, it's measures and countermeasures... forever.

Digital

During the last decade of this century we'll see most communication, from audio broadcasts to television, convert from analog techniques to digital. Digital communication involves a form of scrambling (the digital format is itself a code), and that makes it subject to legislation that prohibits unauthorized descrambling of coded communication. While it will continue to be almost impossible to detect and prosecute those who choose to violate such laws, it may eventually become possible to restrict the sale of equipment that helps commit such crimes. Of course, that practically guarantees an underground market similar to that in the cable and satellite descrambler market today.

Digital will, however, elevate scanning difficulty by at least an order of magnitude, and will substantially increase the cost of the hardware required. Technologies, companies, and even new language will grow to address the gray market needs of the dedicated.

Scrambling based on the Digital Encryption System (DES), however, will probably remain secure against all but the most dedicated eavesdropper. Of course, satellite descramblers depend upon DES, and they've been thoroughly compromised by the "chip" business, fully explained in *The Television Gray Market* – see Bibliography.

There will be many new targets for the scannist, and there is little hope that legislation will keep up with the market and evolving technology, unless we begin passing generic laws of some sort. Historically, we've been reluctant to take such drastic measures unless there was some sort of national threat.

The fastest-growing segment of the electronic industry is Personal Communications, which includes cellular, paging, security, cordless telephones, medical radio, industrial control and monitoring systems, VSATs (very small aperture terminals), wireless local area networks (LANs), wide area networks (WANs), and metropolitan area networks (MANs), plus a host of other wireless and mostly digital applications. Most will be spread spectrum or otherwise digital, and others will be non-digital narrowband FM, but *all* will be scanned in one way or the other.

Legislation against scanning is a bit like forbidding evil thoughts.

New Technology

Assuming our laws allow them, future scanners will probably have the following characteristics:

Faster scanning, to provide very rapid coverage of a chosen frequency range. This will be facilitated by logic (digital devices, including the microprocessor) that can be clocked at ever increasing rates. Such logic exists today, usually in gallium arsenide (faster than silicon or a speeding bullet), but it's too expensive for most consumer applications. Speed will also be enhanced as frequency synthesizer technology evolves.

Better spectral purity, because of improved frequency synthesizer technology, thus increasing range and overall performance.

Longer battery life (and faster re-charging) for portables, plus enhanced performance as battery drain becomes less important and circuitry can be added to do more for the listener.

More memory, to allow storage of complex algorithms and frequencies. The more memory, the more clever the software can be.

Greater processing speed, thus allowing the computer to "think" more about the signal and spectrum activity, and to scan faster.

Active antennas, thus improving sensitivity and providing directionality as well. Scanners today can detect a signal's direction only by use of a directional antenna.

Built-in enforcement of laws whereby a critical integrated circuit contains an inherent characteristic that prevents lawbreaking or modification. Such a subcircuit might examine the received signal and block all those that are "de-authorized" by legislation. That strategy is possible even today, but would involve expense that the consumer would not tolerate, so the industry has resisted such steps.

Digital signal processing (DSP) at ever higher frequencies will push scanners, and perhaps all radios, closer and closer to all-digital solutions. Like many radios of the future, the scanner will eventually consist of an antenna, an amplifier, an analog-to-digital converter, and a microprocessor to manage and manipulate the digital signal and the information contained by it, with results that are unimaginable today. DSP is unquestionably the wave of the future in all communication and entertainment systems, only incidentally including scanners.

Speech recognition, which will permit a scanner to "turn on" when it senses selected spoken words within the transmissions being scanned.

Multiple functions, including paging, Global Positioning System (GPS), AM/FM reception, recording/playback on solid media (digital memory), and many more that we cannot today define. On the other hand, once they're defined, they probably *will* happen.

Features will abound, including several exotic derivatives of CTCSS (a term that denotes tone signaling), smart squelch with different categories of differentiation between noise and meaningful material, multiple priority channels, intelligent interpretation of spoken words, and much more that is beyond today's imagination. Even Spock will be satisfied.

Incredible programmability, with each unit beginning with a blank page of operating characteristics. The consumer will quickly run through menu-driven computer-generated queries that define the operating system, frequencies and bands of interest, those to be avoided, delays, and more, virtually all by speech recognition.

This is all speculation, and as such it's probably wrong. We can be sure that (as usual) the future holds surprises, in this field as it does in all of them. Just as no meteorologist can predict whether a butterfly in Singapore will eventually cause a tornado in Oklahoma, and an army of PhD economists in Washington cannot do more than guess at tomorrow, we cannot predict the future for the scanner hobby except to believe that there will be one.

And it will continue to be very, very interesting.

Shakespeare
Henry IV
(1598)

**"It is the disease of *not*
listening that troubles me..."**

ADDENDA

GLOSSARY

APPENDICES
1. Voice Codes:
   APCO Ten-Codes
   Phonetic Alphabets
2. Clubs and Organizations

BIBLIOGRAPHY
1. Periodicals
2. Books
3. Book Publishers
4. Lists, Directories, & Search Services

INDEX

Samuel Johnson
(1751)
Essay: *The Rambler*

**"Curiosity is one of the
permanent and certain
characteristics of the
vigorous mind."**

GLOSSARY

Every profession, and every hobby, has its own language. Scanning & Short Wave Listening is no exception in that regard, and its language is interesting and descriptive. Because the practice of the hobby is communication by radio, changes in the "lingo" sometimes evolve just as quickly as information travels. This glossary is not comprehensive, but was prepared to reduce the frustration of the newcomer as he enters the hobby and begins developing his own vocabulary.

1.25-meter   Ham band from 222-225 MHz.

2-meter   Ham band from 144-148 MHz.

6-meter   Ham band from 50-54 MHz.

75-cm   Ham band from 420-450 MHz.

Active antenna   Combined antenna and preamplifier. If the preamplifier is good enough (GaAs FET, for instance) it may improve performance.

AM   Amplitude Modulation. Also, refers to the entertainment broadcast band from 550-1650 kHz.

Amateur radio   The hobby, defined by the FCC. Don't let the name deceive: this hobby is comprised of unpaid professionals, and has amassed a pool of knowledge available nowhere else.

Amplifier   Active circuit element that increases power, voltage, or both of an electrical signal.

Amplitude   Voltage or power of an electrical signal. Each has it own units of measurement.

Analog	Data storage and transmission techniques where *any* value whatever can be defined. Generally, this is the real world, so human perception and processing is in analog form.
Angels	Military term for altitude in tens of thousands of feet, as in "Angels 40," which means 40,000'.
Approach	Path of an aircraft from final navigation point to touchdown.
Approach control	Traffic management facility that controls aircraft approaching an airport for landing.
ARTCC	Air Route Traffic Control Center
ASIC	Application-Specific-Integrated-Circuit, which embodies proprietary circuitry in one device. Available only to the "owner," and not usually available over the counter.
ATC	Air Traffic Control. Primary enroute air traffic management agency, with area centers linked by radio, landline, and computer.
Aviation band	118-137 MHz.(military is 225-400 MHz)
Bank	Array of frequencies in a scanner, usually associated with one of the number keys.
Bearing	Compass direction between two points, used for angulated location-finding of radio signal emitters.

Bingo	In military aviation, return to base. Alternatively, to shift to a backup frequency, as in "go to channel white, bingo red."
Birdie	Signal generated within or by the tuning mechanism that deceives the radio into stopping the scan.
BNC	Common antenna connector, twist to lock.
Bogie	Unidentified aircraft spotted by one of the good guys, or by ground radar. "You have a bogie at two o'clock, ten miles."
Broadcast band	550 - 1650 kHz, amplitude modulation.
Bulletin Board	Computer+phone+software, permits modem-equipped callers to leave and receive messages, load and donate software, etc.
Capacitor	Reactive circuit component that stores energy, or passes an alternating signal while blocking a non-alternating one.
Cellular	Mobile telephony, where the portable radio communicates with radios each commanding a "cell," with cells interconnected with the plain old telephone system through the Mobile Telephone Switching Office (MTSO)
Cellular band	869-894 MHz.
Center	See ARTCC
Clearance	The instructions given to a pilot as to route and altitude profile from departure to destination.

Coaxial cable ("coax")   Cable consisting of outer insulating sheath, flexible wire or foil layer, another insulating layer, and a central wire conductor. Used for low-loss radio frequency connections, typically with BNC connectors.

Communications Act of 1934   Legislation designed to give the FCC control over the airwaves, and to limit certain types of activity by requiring licensing and protocol compliance. Amended in 1989 to raise forfeiture limits that the FCC can impose.

Compressive receiver   A sort of scanner, whereby a large segment of the spectrum is received and examined for signals.

Counter   Electrical tool that can receive a signal and display its frequency.

Crystal   Common frequency selection or tuning mechanism for single frequency or few-frequency radios.

CT-2   Second generation cordless phones, usually with bitstream or spread spectrum digital encoding of voice.

CW   Continuous Wave, or Carrier Wave, without modulation.

Data logger   Extended-operation tape recorder, set up to turn on when signals are detected by the scanning receiver. After recording that communication, the unit turns off and awaits the next one.

Decibel	Or "dB." Unit of power ratio, expressed logarithmically. Times 2 = 6dB, times 10 = 20dB, etc.
Delay	Scanner function, usually selectable, whereby the unit remains on frequency for perhaps two seconds following reception of a signal, thus permitting reception of the response to it.
Demodulator	Receiver circuitry that extracts information from a radio signal.
Departure control	Traffic management facility that controls aircraft after takeoff but still in the local control area.
Digital	Data coding, storage, and transmission technique whereby information is reduced to expressions constructed using only two symbols (for convenience, "1" and "0," though could just as easily be "on" and "off," or "green" and "Pontiac." A digital signal can only express a value in discrete increments, while an analog signal can express any value. A convenience, used in computing.
Diode	Circuit element that permits electricity to flow in only one direction.
Dipole	The simplest directional antenna. A TV "rabbit ears" antenna is a dipole.
Direct-frequency tuning	Enables tuning a frequency by literally entering the numbers that define it.
Directional	Antenna with greater sensitivity in some directions and less in others.

Dispatch

Refers to channels and functions that control vehicle movement.

Doppler

Phenomenon whereby a frequency appears to change as the transmitter and receiver approach or separate.

Down-converter

Accessory that receives (typically) 800-900 MHz cellular and public service signals and converts them to frequencies tunable by a scanner that otherwise would not be able to operate in that band.

DTMF

Dual Tone Multi Frequency: the audio heard when a phone key is touched. Made up of two specific tones, with each pair corresponding to a number or function.

Duplex

In two-way communications, duplex permits both sides to speak simultaneously, as over the common telephone.

DVM

Digital Volt Meter. Perhaps the most useful single tool in electronics. Measures AC/DC voltage, current, resistance, and several do capacitance as well.

DX

In amateur radio, "extended distance."

Dynamic range

Signal strength limits between which the unit will operate satisfactorily without malfunction, and refers to range between weakest signal detectable and strongest signal that will not produce overwhelming intermodulation and distortion.

ECPA   Electronic Communications Privacy Act of 1986, designed to prevent eavesdropping upon cellular and cordless phones, though the latter coverage was reversed by the Supreme Court.

ELF   Extremely Low Frequency band, from 0 Hz to 6 kHz, usually used by the Navy for submarine communications.

Ether ("æther")   That medium through which radio signals travel.

Eyeball   In aviation, visual contact.

Femto-   Almost none at all. *Teeny!* Would you believe a quadrillionth?

Fireground   Refers to the fire scene.

Flare   Raising the nose just before touchdown to cushion the landing (Navy/Marine). It's called "round out" in the USAF.

Flight Level   Altitude in tens of thousands of feet (civilian).

FM   Frequency Modulation. Also, a term used to refer to the high-fidelity broadcast band from 88-108 MHz.

Frequency   Specific point (periodicity of vibration, in a way) in the electromagnetic spectrum.

Frequency steps ("resolution")   Tuning increments of a digital tuning device such as a synthesizer.

Fundamental   A single-frequency signal, with neither harmonics nor subharmonics.

Gain
Increase in effective radiated power (in a transmitter) or receptivity (in a receiver), when compared to omnidirectional antennas or passive circuitry.

Ground Control
Airport facility that controls and coordinates taxiing aircraft and vehicle traffic.

Ground plane
Conductive or electrically reflective surface under an antenna to serve as an artificial earth ground.

Guard
A term that denotes emergency aviation frequencies, always monitored by all aircraft and by all ground control agencies. VHF – 121.5 MHz, UHF – 243.0 MHz.

Guard band
"Margin" next to a selected frequency. Extra room to prevent adjacent channel interference.

Handheld
Small, portable, battery-operated device.

Harmonic
A frequency exactly twice that of a "fundamental" frequency, which appears and disappears as the fundamental is switched on and off, thereby proving a relationship.

HF
High frequency band, most-used by hams.

IC
Integrated circuit.

IFR
Instrument Flight Rules, where a ground facility maintains clearance between the aircraft and ground, other aircraft, etc. Under IFR, the pilot often flies "blind."

Image   When the received signal is downconverted to an intermediate frequency, an "image" is generated at twice that number.

Impedance   Resistance, to an AC (typically radio frequency) signal.

Inductor   Reactive circuit component that passes a non-alternating signal while blocking an alternating one.

Inquiry   Refers to channels used for questions.

Inter-modulation ("Intermod")   The new frequency products (sum and difference) when two signals are mixed. These are new frequencies, detected as radio signals, and therefore generate birdies and fool the scan mechanism into stopping.

Keylock   Switch or lever that prevents inadvertent keyboard entries.

Lambda (λ) Greek symbol used to denote wavelength.

LF   Low Frequency band, from 30-300 kHz.

Lockout   Scanner feature that skips past undesired frequencies or channels.

Logic   Integrated circuits that comprise the thinking mechanisms of a computer.

Loop   Simple directional antenna.

LSB   Lower Sideband: A derivative of SSB where the carrier and upper sideband are suppressed, and information is carried only on the lower sideband.

Mayday	In aviation, EMERGENCY.
MF	Medium Frequency band, from 300 kHz to 3 MHz. Includes the common "AM" broadcast band.
Micro-	One millionth.
Micro-processor	Also μP: the computer's brain, consisting of an array of logic.
Microwave	Anything above 2 GHz in frequency.
Mobile	Intended for vehicle installation.
Modulation	The means by which information is impressed on a fundamental radio signal.
Modulator	Transmitter circuitry that imposes information upon a radio signal.
Monitor	Scanner function that stores a frequency of interest found by searching.
MTSO	Mobile Telephone Switching Office, which connects the cellular cell radios with the telephone system.
Nano-	One billionth.
NOAA	National Oceanographic and Atmospheric Administration: weather alert service.
Non-volatile	Refers to memory. Non-volatile retains information even when power is removed.

NSA	National Security Agency, at Fort George G. Meade, between Baltimore and Washington DC.
Octave	The range from any frequency to exactly twice that frequency, both in audio and in radio.
Omni-directional	Refers to antennas with equal sensitivity to signals from every direction.
Omnibus	Originally called the Omnibus Crime Control and Safe Streets Act of 1968, attempts to control eavesdropping by establishing penalties.
Oscilloscope	Second most important test instrument, displays waveforms in the *time domain*. That is, it shows relationships between time and amplitude.
Pan	In aviation, deferred emergency (seldom heard).
PCN	Personal Communications Networks (also Personal Communication Systems), from 902-928 MHz, also in low-microwave.
PCS	Personal Communication Systems (see PCN)
Phase noise	Specification that identifies the quality of the signal used to tune the radio/scanner, hence the selectivity.
Pico-	One trillionth.
Pirate radio	Unlicensed, unauthorized broadcasts, usually political or entertainment, often from ships at sea, or across the border..

POTS

Plain Old Telephone Service.

Preamplifier

Also "booster," amplifies radio signals and noise, together.

Priority

Scanner function. When a channel is designated "priority," every few seconds the unit returns to check for transmissions on that frequency.

Public safety

Police, fire, ambulance, rescue, disaster relief, and other emergency services.

Quartz ("quartz-lock")

Refers to circuit that is tuned using a frequency synthesizer, based upon a crystal reference, rather than a reactive passive circuit.

Range

That distance over which a signal can be received by a given set of factors, such as transmitter amplifier, antenna gain at each end, and receiver sensitivity.

Repeater

A combined receiver and transmitter, on two frequencies, whereby a weak signal is received and retransmitted with more power, thus increasing range.

Resistance

Constriction of flow, requiring that the electrical signal give up power to pass.

Resistor

In a circuit, a constriction through which electrical energy passes while giving up some of its energy.

Rosin core   The *right* solder to use when working on anything electronic. "Acid core" solder is for plumbing, and should be avoided at all costs.

RTTY   Radio teletype, using the 5-bit International Telegraph Alphabet #2 (ITA2). Usually ham communication below 30 MHz, and at low data rates.

Rubber Duck   Rubber, flexible antenna with which most handheld scanners are equipped.

Scanner   Radio capable of automatically searching for transmissions either through the spectrum, or through a predefined series of frequencies.

Scannist   Scanner enthusiast/hobbyist. Hopefully will become widely used.

Scrambler   Means by which a transmission is encoded specifically to avoid eavesdropping or unauthorized reception.

Search   Scanner sweeping between preset high and low frequency limits.

Selectivity   Scanner specification that indicates the unit's ability to discriminate between two signals close together.

Simplex   In two-way communications, simplex is a single path requiring that only one "end" speak at a time.

Skip   Radio energy propagation that "bounces" off of an ionized atmospheric layer, increasing range beyond the expected line of sight.

SMA   Microwave connector for high frequency use.

SMB   Microwave connectors that operate over a broader frequency range than the less expensive BNC.

Spectrum   The delimited range of *anything,* such as color, hardness, etc. In the context of this book, refers to a range of periodicities, or frequencies, available to radio and other electromagnetic devices.

Spectrum analyzer   Instrument that depicts some segment of the spectrum and the signals detected within it. Display is in the *frequency domain*, and shows the relationship between energy and frequency.

Spread Spectrum   Modulation technique whereby an integrated digital code is used to distinguish the signal from other energy, allowing it to be extracted even from noise.

Spurs (spurious)   Uncommanded discrete signals generated by the tuning mechanism, sometimes heard as "birdies."

Squelch   Scanner function that reduces sensitivity to all but signals of desired amplitude.

SSB   Single Sideband. A modulation strategy whereby the primary carrier is suppressed.

SSTV   Slow-scan TV, a hobbyist method of transmitting video images using relatively low data rates and narrow bandwidth.

State	In aviation, refers to fuel level, as in "What's your state?"
SWAT	Special Weapons and Tactics.
Synthesizer ("frequency synthesizer")	Tuning device that can generate many precision frequencies, each locked to a quartz crystal reference.
TA (Tango Alpha)	In aviation, a traffic advisory.
Tabletop	Intended for base station operation.
Telephone Disclosure and Dispute Resolution Act	1992 federal legislation, originally proposed to control abuse of 900-numbers, but with addenda that cause the FCC to restrict cellular-capable scanners.
Trunking system	Public safety and industrial radio network where a bank of frequencies is available to all stations, and traffic is controlled by a computer that communicates over a management channel.
TVRO	TeleVision Receive-Only ("satellite receiver").
USB	Upper Sideband. One derivative of SSB, where the carrier and the lower sideband are suppressed, and information is carried only on the upper sideband.
UTC	Universal Coordinated Time (World Time), once called Greenwich Mean Time. And it *is* U T C, not U C T.

VFR

Visual Flight Rules, used when weather permits. Under VFR, the pilot maintains his/her own clearance from other aircraft, terrain, etc.

VHF

Very High Frequency band, usually defined as 30-174 MHz.

Victor

Term used to identify standard numbered airways, as in "cleared to X-ray on Victor 33." Rather like an interstate highway system at 36,000'.

VLF

Very Low Frequency band, from 6 to 30 kHz.

Volatile

Refers to memory. Volatile memory loses information when power is removed.

VSAT

Very Small Aperture Terminal, refers to either satellite or terrestrial (point-to-point) telecommunication terminals.

Wavelength

Distance between two corresponding points in a signal's repetitive waveform

Whip

Simple antenna. A straight piece of wire, usually of a calculated length.

Yagi

Complex directional antenna, with long, narrow, high-gain pattern. A rooftop TV antenna is somewhat like a yagi.

Zulu

Refers to Greenwich Mean Time (at Greenwich, England), also "UTC."

APPENDIX 1: Voice Codes

**Ten-Codes endorsed by the Associated Public
Safety Communication Officers (APCO)**

TEN—

1 *Cannot understand your message*
2 *Your signal is good*
3 *Stop transmitting*
4 *Message received ("OK")*
5 *Relay information to* _____
6 *Station is busy*
7 *Out of service*
8 *In service*
9 *Repeat last message*
10 *Negative*
11 _____ *in service*
12 *Stand by*
13 *Report* _____ *conditions*
14 *Information*
15 *Message delivered*
16 *Reply to message*
17 *Enroute*
18 *Urgent*
19 *Contact* _____
20 *Unit location*
21 *Call* _____ *by telephone*
22 *Cancel last message*
23 *Arrived at scene*
24 *Assignment completed*
25 *Meet* _____
26 *Estimated time of arrival is* _____
27 *Request information on license*
28 *Request vehicle registration information*
29 *Check records*
30 *Use caution*
31 *Pick up*
32 *Units requested*
33 *Emergency! Clear the air*
34 *Correct time*

APPENDIX 1: Voice Codes
Phonetic Alphabets

At least two phonetic alphabets are in common use. In about 1959, the military shifted from *ABLE-BAKER* to *ALPHA-BRAVO*, and most of the world followed, including law enforcement. The secondary alphabet, shown in lighter type, is used in many areas, and often interchangeably with the primary alphabet.

ALPHA	ADAM	**NOVEMBER**	NORA
BRAVO	BOY	**OSCAR**	OCEAN
CHARLIE	CHARLES	**PAPA**	PAUL
DELTA	DAVID	**QUEBEC**	QUEEN
ECHO	EDWARD	**ROMEO**	ROBERT
FOXTROT	FRANK	**SIERRA**	SAM
GOLF	GEORGE	**TANGO**	TOM
HOTEL	HENRY	**UNIFORM**	UNION
INDIA	IDA	**VICTOR**	VICTOR
JULIET	JOHN	**WHISKEY**	WILLIAM
KILO	KING	**X-RAY**	X-RAY
LIMA	LINCOLN	**YANKEE**	YOUNG
MIKE	MARY	**ZULU**	ZEBRA

When communicating numbers, most professionals use *ZERO* (not "OH") and *NINER* (to prevent confusion with FIVE).

When listening to vehicles in action, *ONE* is routine, *TWO* is urgent but without lights/siren, and *THREE* is urgent *with* lights/siren. Though every jurisdiction and area has its own local language, for obvious reasons a few terms are used reasonably consistently nearly everywhere in the U. S.:

DL	Driver license	**DOA**	Dead on arrival
DOB	Date of birth	**ETA**	Estimated time of arrival
NCIC	National Crime Information Center	**VIN**	Vehicle ID number

APPENDIX 2
Scanner/SWL Clubs & Organizations

All Ohio Scanner Club
50 Villa Road
Springfield, OH 45503
American Scannergram

American SW Listener's Club
16182 Ballad Lane
Huntington Beach, CA 92649

**Association of Clandestine
Enthusiasts**
POB 11201
Shawnee Mission, KS 66207
The A.C.E.

**Association of DX Reporters
(ADRX), MW, LW, and SWBC**
7008 Plymouth Road
Baltimore, MD 21208
DX Reporter

**Association of Manitoba DX'ers
(AMANDX)**
30 Beacontree Bay
Winnipeg, Manitoba R2N 2X9 Canada

Bay Area Scanner Enthusiasts
4718 Meridian Ave., #265
San Jose, CA 95118
Listening Post

Bayonne Emergency Radio Network
POB 1203
Bayonne, NJ 07002

Bearcat Radio Club
PO Box 291918
Kettering, OH 45429
National Scanning Report

Boston Area DXers
9 Sterling Street
Andover, MA 01810

**British Columbia Shortwave
Listening Club (BCDX)**
Box 500 2245 Eton St
Vancouver, BC V5l 1C9
Canada
LOGJAM

Central Indiana Shortwave Club
2517 E. DePauw Road
Indianapolis, IN 46227
Shortwave Oddities

Canadian Int'l DX Club
79 Kipps Street
Greenfield Park, Quebec Canada
J4V 3B1
The Messenger

Capitol Hill Monitors
6912 Prince Georges Ave
Takoma Park, MD 20912-5414
Scannerbands
BBS 703-207-9622

Central Florida Listeners Club
956 Woodrose Court
Altamonte Springs, FL 32714

**Chicago Area Radio Monitoring
Association**
6536 N. Francisco
Chicago, IL 60645

Communications Research Group
122 Greenbrier Drive
Sun Prairie, WI 53590

DelcoMania
PO Box 126
Licroft, NJ 07738
DelcoMania

Fire Net
Box 1307
Culver City, CA 90232

Global DX Club
PO Box 1176
Pinson, AL 35126-1176
Radio Waves

**Houston Area Scanners &
Monitoring Club**
909 Michael
Alvin, TX 77511

**Hudson Valley Monitors Association
(HVMA)**
PO Box 706
Highland, NY 12528
The Hudson Valley Monitor

International 11 Meter Alliance
Rt 1 Box 187-A
Whitney, TX 76692
Public safety

International Radio Club of America
PO Box 70223
Riverside, CA 92513
DX Monitor

Long Island Sounds
2134 Decker Ave.
North Merrick, NY 11566

LongWave Club of America
45 Wildflower Road
Levittown, PA 19057
The Lowdown

**Memphis Area Shortwave Hobbyists
(MASH)**
PO Box 3888
Memphis, TN 38173

Metro Radio System
PO Box 26
Newton Highlands, MD 02161
MRS Newsletter

Michigan Area Radio Enthusiasts
PO Box 81621
Rochester, MI 48308
Great Lakes Monitor

Minnesota DX Club
PO Box 10703
White Bear Lake, MN 55110
MDXC Newsletter

Monitoring the Long Island Sounds
2134 Decker Ave
North Merrick, NY 11566
Primarily Scanner

MONIX
(Cincinnati, Dayton area)
7917 3rd Street
West Chester, OH 45069

Mountains News Net
PO Box 621124
Littleton, CO 80162-1124
Public safety notification
Mile High Pages

National Radio Club
PO Box 5711
Topeka, KS 66605
DX News

National Radio Club-DX
PO Box 164
Mannsville, NY 13661-0164

North American SW Association
45 Wildflower Lane,
Levittown, PA 19057
The NASWA Journal

North Central Texas SWL Club
1830 Wildwood Drive
Grand Prairie, TX 75050

Northeast Ohio SWL/DXers
PO Box 652
Westlake, OH 44145

NYC Radio Fre(ak)qs
199 Barnard Ave.
Staten Island, NY 10307

Ontario DX Association
PO Box 161, Station A
Willowdale, Ontario M2N5S8 Canada
DX Ontario

Pacific NW BC/DX Club
9705 Mary NW
Seattle, WA 98117
PNBCDXC Newsletter

Pitt Co SW/ Scanner Listeners
Rt 1 Box 276 Sumrell Road
Ayden, NC 28513-9715
The DX Listener

Puna DX Club
P O Box 596
Keaau, HI 96749
Puna, HI; SW and MW

Radio Monitors of Maryland
PO Box 394
Hampstead, MD 21074
*Radio Monitors Newsletter of
Maryland*

RCMA
PO Box 542
Silverado, CA 92676
Scanner Journal

Regional Communications Network
PO Box 83M
Carlstadt, NJ 07072

**Rocky Mountain Monitoring
Enthusiasts**
11391 Main Range Trail
Littleton, CO 80127

Rocky Mountain Radio Listeners
4131 S. Andes Way
Aurora, CO 80013

Scanning Wisconsin
67W 17912 Pearl Dr.
Muskego, WI 53150
Scanning Wisconsin

Southern California Area DXers
3809 Rose Ave.
Long Beach, CA 90807

SPEEDX
PO Box 196
DuBois, PA 15801
SPEEDX

Susquehanna County Scanner Club
PO Box 23, Prospect St.
Montrose, PA 18801

Toledo Area Radio Enthusiasts
6629 Sue Lane
Maumee, OH 43537

Triangle Area Scanner/SWL
PO Box 28587
Raleigh, NC 27611

Wasatch Scanner Club
2872 West 7140 South
West Jordan, UT 84084
Newsletter/directory

World DX Club
Richard D'Angelo
2216 Burkey Drive
Wyomissing, PA 19610

**Worldwide TV/FM Dxers
Association (WTFDA)**
PO Box 514
Buffalo, NY 14205-0514
VHF/UHF Digest

Heraclitus
(460 BC)
"It is wise to listen..."

BIBLIOGRAPHY

PERIODICALS

9-1-1 Magazine
18201 Weston Place
Tustin, CA 92680
714-544-7776

Emergency services, and there's no doubt about its mission. A gold mine for Public Safety wannabes and professionals.

Communications Quarterly
PO Box 465
Barrington, NH 03825

Theoretical/technical issues in amateur radio.

CQ, The Radio Amateur's Journal
CQ Communications, Inc.
76 North Broadway
Hicksville, NY 11801-2953

Though slanted toward the ham, this magazine presents a wealth of technical and spectrum data useful to serious scanner enthusiasts, too.

Dispatch
2945 David Lane
Medford, OR 97504

Professional magazine about dispatching, including technology, etc.

DX Listening Digest
Box 1684
Enid, OK 73702

Specialized, long distance listening.

Electronics Now
500-B Bi-County Blvd.
Farmingdale, NY 11735
516-293-3000

Great how-to, and a gold mine for the electronics hobbyist as much for the advertising as for the often irreverent & useful articles. A broad variety of topics, not just scanning.

Global DX Newsletter
Box 1176
Pinson, AL 35126

Sample $1

Intercepts Newsletter
6303 Cornell
Amarillo, TX 79109

Focused upon military monitoring.

Monitoring Times
PO Box 98
Brasstown, NC 28902-0098
704-837-9200

For the serious SWL and scanner enthusiast, in that order. Excellent editorial material, and frequencies seldom printed elsewhere.

National Scanning Report
PO Box 291918
Kettering, OH 45429
800-423-1331

For the devotee, with data seldom published elsewhere. Serious and useful.

Popular Communications
by CQ Communications, Inc.
76 North Broadway
Hicksville, NY 11801-2953

Most widely read magazine among scanner hobbyists. Excellent editorial content. Interesting and useful ads. Mandatory!

QST
ARRL
225 Main St.
Newington, CT 06111
203-666-1541

Focused on amateur radio, includes many excellent tidbits of technical and product information useful in scanner hobby.

Radioscan Magazine:
The International
Amateur Radio Digest
8250 NW 27 Street, Suite 301
Miami, FL 33122

Primarily for the SWL, but with good technical & spectrum usage data.

Scanner Journal
PO Box 542
Silverado, CA 92676

Deputies in every state help ensure current data. Great coverage of railroads and other special interests. Formerly *RCMA Journal*.

U. S. Radiosport
PO Box 190176
St. Louis, MO 63119

Bi-monthly, for DXers and contesters. Serious SWL.

U. S. Scanner News
706 W. 43rd St.
Vancouver, WA 98660

Scanners, VHF/UHF.

World Scanner Report
Bill Cheek/CommTronics
PO Box 262478
San Diego, CA 92196

Professional, focused, and technically excellent. A great help, and it provides the best "how-to" data in the scanner world.

BOOKS

Only a few titles are listed. Some of them do not appear in *Books in Print*, are retailed by multiple vendors, and advertisements for them often exclude the author's name and the publication date. For those reasons, conventional bibliographical format is ignored and the list below includes only enough information to identify the book, plus a typical price and at least one retail source (which *may* be the publisher). A few pertinent publishers are identified in the section that follows, and some are listed with no books shown, so readers are encouraged to call or write for catalogs.

Anon. Air Waves, $18, DX.

Anon. Cop Talk, $20, DX, CRB.

Anon. Fire Call, $20, DX.

Anon. Realistic Modification Instruction Manual, and Eavesdropping for Fun and Profit (combined). $14, Starlite.

Anon. Scanner Hackers Bible, $34, Telecode.

Anon. Scanner Repair and Modification Manual, $35, Thomas.

ARRL. First Steps in Radio. $5. Tutorial, includes basic electronics plus theory required for licensing. ARRL.

ARRL. Repeater Directory. $6. Lists more than 18,000 VHF/UHF repeaters in North America. ARRL.

ARRL. The ARRL Handbook (70th edition!!). $25. ARRL.

ARRL. The ARRL Radio Buyer's Sourcebook. $15. Reviews and commentary, modification data. ARRL.

ARRL. Understanding Basic Electronics. $17. An excellent textbook – starts with basic math, addresses key components and circuitry, and ends at the system level. ARRL.

Bell, Bob. Listening In To Aircraft Communications, $25, Air Band. Written for Australia, and publisher claims it's applicable worldwide.

Bornstein, Howard. Guide to the AR1000, $15, Grove. Applies to a variety of scanners.

Cheek, Bill. Scanner Modification Handbook (Vols 1 and 2), 160 and 220 pages respectively, softcover, $18 each, CRB. Well written and technically sound books on popular scanner hardware, with many modifications and improvements.

Cheek, Bill. The Ultimate Scanner (Cheek[3]). 260 pages, softcover, large format, $24.95, Index. This comprehensive new book begins where his previous two bestsellers end, adding channels, functionality, convenience, and performance. A mandatory resource for any scannist.

Creech, Kenneth. Electronic Media Law and Regulation, $34, Focal.

Curtis, Anthony R. Monitoring NASA Communications, $15, CRB.

Eisenson, Henry. TravelScan; Good Frequencies Across America, $7.95, Index. The pocket guide for scannists on the go, with the most popular frequencies of the top 100 cities in the U. S. and Canada, plus state information, recreation, highway data, and much more.

Eisenson, Henry. The Television Gray Market, $24, Index. 168pp. On the underground business of stealing cable and satellite programming.

Flynn, Ed. Understanding ACARS. Manual on the Aircraft Communication Addressing and Reporting System. Universal Radio Research.

Grove, Bob. Receivers and Scanners Pricing Guide. $6, Grove.

Grove, Bob. Scanner and Shortwave Answer Book, $13, Grove. An excellent resource for the experienced and the beginning scanner enthusiast, with hundreds of questions and answers, well illustrated.

Harrington and Cooper. Hidden Signals on Satellite TV, $20, Universal.

Helms, Harry. The Underground Frequency Guide, $11, CRB.

Illman, Paul. The Pilot's Radio Communications Handbook, $17, CRB.

Kneitel, Tom. Air-Scan, 5th Edition, 192p, softcover, $15, CRB. On scanning aviation. More than just frequencies – lots of information, too.

Kneitel, Tom. "Top Secret" Registry of U. S. Government Frequencies, 268p, $22, CRB, Index. Useful and widely-read "lists." Answers many questions about government communication, and creates opportunities to find out how our government works, or doesn't. A good resource.

Kneitel, Tom. Tune In On Telephone Calls, 160p, softcover, $12.95, CRB, Index. Useful if phone calls are your passion. Lists many less obvious "telephones," such as ship-to-shore, air-to-ground, industrial radio-telephone, etc.

Quarantiello, Laura. Citizen's Guide to Scanning, $20, DX, Grove.

Schrein, Norm. EMERGENCY RADIO! Scanning News as it Happens. 214 pages, $14.95, Index. The "how" and "why" we scan, by the President of the Bearcat Radio Club, concentrating on the public service agencies of our country and beyond.

Smith, Winston. Covert Techniques for Intercepting Communications, $10, 160pp, CRB.

Soomre, Ed. The Scanner Listener's Handbook, $15. 130pp, CRB.

Terranella, Frank. Listener's Lawbook, $10. 48pp, Grove.

Thorn, Damien. Cellular Fraud; The Vulnerability of Cellular Technology. 325 pages, $24.95, Index. Authoritative and detailed, by *Nuts and Volts Magazine*'s most popular columnist. Tells *everything* about cellular telephones and related technology.

Van Horn, Larry. Monitoring the Strategic Air Command, $13, DX, CRB.

World Radio TV Handbook. Equipment Buyers Guide, $20. Grove.

PUBLISHERS OF SCANNING BOOKS

Air Band Communications
PO Box 16
George Hall
2198 NSW Australia

One title found, and advertisement asserts it's as useful anywhere as it is "down under."

ARRL
American Radio Relay League
225 Main Street
Newington, CT 06111
203-666-1541

The number one publisher in amateur radio, with many titles helpful to the scanner enthusiast, too. Catalog published in *QST*.

CRB Research Books
PO Box 56
Commack, NY 11725
516-543-9169

A very prolific vendor & publisher in the scanning field, with many titles and frequency lists.

Doyle Communications
Route 8, Box 18
Lake Pleasant, NY 12108
518-548-5515

Several titles on scanning, plus frequency lists.

DX Radio Supply
Box 360
Wagontown, PA 19376

Scanning and SWL specialist.

Focal Press
800-366-2665

One title found.

Grove Enterprises
PO Box 98
300 S. Hwy 64 West
Brasstown, NC 28902-0098
704-837-9200

Highly authoritative. Both *publishes* excellent books and *sells* selected titles from other publishers.

High Text Publications
PO Box 1489
Solana Beach, CA 92075
800-247-6553

Shortwave and scanner books and papers.

Index Publishing Group
3368 Governor Drive, St. 273F
San Diego, CA 92122
800-546-6707

Ambitious, professional,
and focused. Modesty
forbids...

ITS, Inc.
2100 "M" Street NW, #120
Washington, DC 21037

Headquarters. Fiche lists.

1270 Fairfield Road
Gettysburg, PA 17325

Field office.

Paladin Press
Box 1307
Boulder, CO 80306
303-443-7250

Huge catalog covers host
of uncommon topics,
including scanning.
Catalog alone is very
interesting reading.

Starlite Mfg. Co.
4424 Clemice Lane
Montgomery, AL 36106

One title found.

Thomas Distributing
128 East Wood
Paris, IL 61944

One title found.

Universal Distributing
4555 Groves Road, Suite 13
Columbus, OH 43232
800-241-8171

Major dealer in scanner
and shortwave radios, and
sells a wide range of
books on the hobby.

Universal Radio Research
6830 Americana Parkway
Reynoldsburg, OH 43086

ACARS manual by
Ed Flynn

FREQUENCY LISTS, DIRECTORIES, SEARCH SERVICES

This section includes resources for frequency information that is of reasonably broad interest. Some agencies publish frequency lists, others microfiche, and a few conduct custom search services for specific areas of interest.

Several state and regional lists are included, but for the most complete and timely "local" information, attend a meeting of an appropriate club or organization in your area.

An extremely comprehensive resource is Bob Grove's annual *FCC DATABASE ON DISK*, and CD-ROM national and state databases are available from Percon.

Aeronautical Frequency Directory
Official Scanner Guide
PO Box 712
Londonderry, NH 03053

Aircraft Frequency Directory
DC Enterprises
3420 Trenary Lane
Colorado Springs, CO 80918

Bellows, G.
(Search service)
PO Box 1239
Charleston, SC 29402

Betty Bearcat Scanner Guide
Bearcat Radio Club
PO Box 291918
Kettering, OH 45429
800-423-1331

Confidential Freq List
Gilfer Shortwave
52 Park Ave.
Park Ridge, NJ 07656
201-391-7887

Directory of North American Military Aviation Communications
2d Ed. NE, SE, Centr, W.
Hunterdon Aero Publishers
PO Box 754
Flemington, NJ 08822
800-542-SCAN

FCC Master Frequency Database (on CDROM)
ALL freq allocations for the U.S., with data on each freq. State databases, also.
PERCON Corp
4906 Maple Springs/Ellery Road
Bemus Point, NY 14712
716-386-6015

*Frequency Assignment
Master File*
Artsci Publications
PO Box 1848
Burbank, CA 91507
818-843-4080

Grove/FCC Database
[A truly comprehensive
source]
Grove Enterprises
300 S. Hwy 64 West
Brasstown, NC 28902-0098
800-438-8155

*Grove's Sports &
Entertainment Frequency
Directory*
CRB, Grove

*How-To Report on Cellular
Listening*
M. D. Koon
4111 S. Alaska St.
Tacoma, WA 98408

*International Callsign
Directory*
Grove Enterprises
140 Dog Branch Road
Brasstown, NC 28902
800-438-8155

*Klingenfuss Guide to
Utilities*
Grove

M Street Radio Directory
Grove

Monitor America
Scanner Master
PO Box 428
Newton Highlands, MA
02161
800-722-6701

Monitoring the Military
Daryll Symington
Grove

*National Directory of
Survival Radio Frequencies*
CRB

*National Highway Patrol
Frequency Handbook*
CRB

Ohio Scanner Pocket Guide
CRB

Police Call (9 volumes)
Hollins Radio Data
POB 35002
Los Angeles, CA 90035
(also at any Radio Shack)

*Professional Florida
Statewide Scanner Guide*
CRB

Scan Air
Heald
6886 Jefferson St.
North Branch, MI
313-688-3952

Scan Rail
Heald

Scanner Master State & Area Guides
MA, ME, NH, VT, CT, NY METRO, NJ, PHILA, & S. NJ, VA, DC, FL, IL
Pocket Guides
Scanner Master
PO Box 428
Newton Highlands, MA 02161
800-722-6701

Secret Frequency Guide
Panaxis Productions
PO Box 130
Paradise, CA 95969

The Pirate Radio Directory
Tiare Publications
PO Box 493
Lake Geneva, WI 53147

TravelScan
Good Frequencies Across America
Index Publishing Group
1-800-546-6707

US DOD Aeronautical Publications
Aerial Development of New England
PO Box 661
Bangor, ME 04402

US FM Broadcast Directory
FM Atlas
PO Box 336
Esko, MN 55733
218-879-7676

U. S. Maritime Frequency Directory
Grove

World Press Services Frequencies
Grove

SCIENCE FICTION SERIAL
X-Man (1949)

"This Z-band radio detects private conversations even though they're miles away. I dial in the right frequency and I hear everything. You have no secrets from me, Kagor!"

INDEX

...about the Author:

Henry Eisenson has published under several *noms de plumes* and in diverse fields since retiring from the Marine Corps, where he was an aviator and recipient of the Silver Star, two DFCs, and an array of other combat awards. He co-founded and is now president of an ultra high-tech electronics firm, whose aggressive technologies have made contributions to our nation's capabilities in electronic warfare, telecommunication, radar, battlefield simulation, and even personal communication systems.